THE RISE OF YEAST

THE
RISE
OF
YEAST

HOW THE SUGAR FUNGUS
SHAPED CIVILIZATION

NICHOLAS P. MONEY

OXFORD
UNIVERSITY PRESS

OXFORD
UNIVERSITY PRESS

Oxford University Press is a department of the University of Oxford.
It furthers the University's objective of excellence in research, scholarship,
and education by publishing worldwide. Oxford is a registered trade mark of
Oxford University Press in the UK and certain other countries.

Published in the United States of America by Oxford University Press
198 Madison Avenue, New York, NY 10016, United States of America.

Library of Congress Cataloging-in-Publication Data

Names: Money, Nicholas P., author.
Title: The rise of yeast : how the sugar fungus shaped civilization /
Nicholas P. Money.
Description: New York, NY : Oxford University Press, [2018] | Includes
bibliographical references and index.
Identifiers: LCCN 2017022904 | ISBN 9780190270711 (hardback : alk. paper)
Subjects: LCSH: Yeast. | Microorganisms.
Classification: LCC QR151 .M73 2018 | DDC 579—dc23 LC record available at
https://lccn.loc.gov/2017022904

1 3 5 7 9 8 6 4 2
Printed by Edwards Brothers Malloy, United States of America

To Judith, my loving and elegant mother

ACKNOWLEDGMENTS

Contemporary publications on yeast were accessed through the online resources furnished by the Miami University Libraries in Oxford, Ohio. Older studies were located in the outstanding collection of books and manuscripts in the Lloyd Library and Museum in Cincinnati. Information on the Fleischmann family was unearthed in the Cincinnati History Library and Archives. Sincere thanks to my scientific research partner, Mark Fischer, for lending his artistry to many of the illustrations in this book.

CONTENTS

LIST OF ILLUSTRATIONS

Wine is life

—Petronius, *Satyricon*

YEAST IS LIFE!
—Advertisement for Irving's Yeast-Vite Tablets (1925)

1

~

Introduction

Yeasty Basics

This is the story of our ancient co-dependence with yeast, how microbe and man have led each other through history, and how the relationship is blossoming in the twenty-first century. From morning toast to evening wine, yeast is mankind's daily blessing, and our communion with this little fungus is deepening year by year. It has taken us from hunting and gathering to more settled lives as farmers, provided our daily bread, and fed our thirst for wine and beer. Without yeast, earth would be an alcohol-free planet and every loaf of bread would be unleavened. In our time, yeast has become the darling of biotechnology, generating a catalog of life-saving medicines, and billions of gallons of biofuel in the quest to slow climate change.

Yeast, the sugar fungus, has been civilization's invisible partner from the get go (Figure 1). Ten thousand years ago, our ancestors abandoned bush meat and wild fruit in favor of farming animals and cultivating grain. Leaving the forests and grasslands, our appetite for the beer and wine produced by the fungus was a major stimulus for agricultural settlement. The reason was simple: it takes a village to run a brewery or tend a vineyard. Modest consumption of alcohol by the early farmers also helped to cement social ties and foster a sense of community. And as life became more predictable

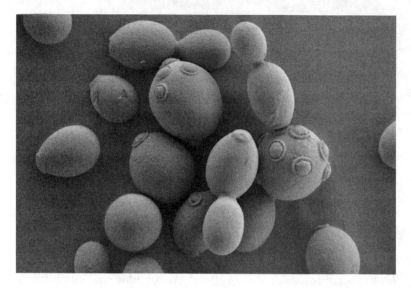

Fig. 1. Scanning electron micrograph of the sugar fungus, *Saccharomyces cerevisiae.*

within these farming communities, the technologies of brewing and baking evolved.

We knew what yeast did long before we had a clue about what it was. Six thousand years ago, Sumerian brewers attributed the fermentation process to the goddess Ninkasi, and many other deities were credited with brewing alcohol in antiquity. In his famous dictionary, published in 1755, Dr. Johnson defined *yest* as, "The foam, spume, or flower of beer in fermentation," and cross-listed with the synonym *barm* as, "Yeast; the ferment put into drink to make it work."[1] The replacement of these original religious and industrial concepts of yeast with the biological description of a microbe happened very slowly. Yeast cells were among the first microorganisms seen after the invention of the microscope in the seventeenth century. Anton van Leeuwenhoek examined them in drops of beer in 1680, although he did not consider that these tiny globules were alive. Chemists, including the French scientist Antoine Lavoisier, studied the fermentation process in winemaking in the next century, and investigators who saw yeast with their primitive microscopes concluded that it was produced by the fermentation process rather than vice versa. With no knowledge of germs as living entities that could carry out chemical transformations or cause disease, there was no reason to view yeast cells as anything worthy of further scrutiny.

Recognition that yeast was a living thing that produced alcohol came in nineteenth century. Looking at the fungus in beer, French botanist Jean Baptiste Henri Joseph Desmazières named it *Mycoderma cervisiae* (*cervisia* is Latin for beer), and called the same organism *Mycoderma vini* when he saw it in wine.[2] German biologist Theodor Schwann called yeast *Zuckerpilz*—the sugar fungus or sugar mushroom—and his colleague Franz Meyen provided the "modern" Latin name in 1838: *Saccharomyces cerevisiae* (Figure 2).[3]

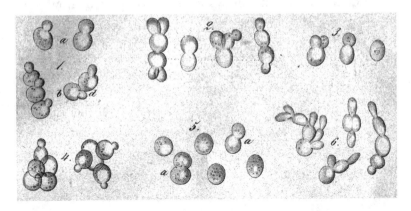

Fig. 2. *Saccharomyces cerevisiae* illustrated by M. Rees in 1870.

The combination of better microscopes and clever experiments on fermentation encouraged the conclusion that the yeast plant, as it became known, was the live agent that produced alcohol in wine and beer.[4] Organic chemists continued to dispute these findings, choosing to believe that the objects described as cells were minerals precipitated from chemical reactions. They thought that alcohol was a product of pure chemistry rather than biochemistry. But with evidence mounting in the 1860s in favor of yeast as the catalyst, Louis Pasteur silenced most of the dissenting voices with a series of brilliant experimental demonstrations.[5] Yeast was proven to be a living entity and recognized as a microorganism of spectacular consequence in human affairs.

There is some Western bias in this assessment of yeast's supremacy. Though the fungus is foundational to human civilization, our nutritional reliance on yeast is concentrated among the descendants of the Roman Empire living in Europe, North Africa, the Middle East, Australasia, and the Americas. Leavened bread has always been less important in Asian and Sub-Saharan African cultures. Alcohol consumption follows a more complex demography. Practitioners of Islam, Mormonism, and many other Christian denominations are teetotal, but at least two billion humans enjoy brewed and distilled beverages today.

According to Pliny, unleavened bread was replaced by loaves raised with yeast after the Third Macedonian War (171–168 BC).[6] Wine production peaked in the Roman Republic at around the same time, and vineyards spread across the conquered territory of Europe to meet the demands of a rapidly growing population. Leavened bread has held its own in the Western diet over the 2,000 years since Pliny's departure and our love of wine and beer is undiminished.

The allure of alcohol is undeniable. If yeast had never evolved, we would have been forced to invent it. Wine and beer change our perception of the moment, can make us glory in life, and be our downfall when absorbed too freely. Alcohol can help us slide from heaven to hell and vice versa. It bathes the central nervous system, acting both as a powerful stimulant and depressant, which explains the effects of different doses ranging from mild euphoria to death. Its legality allows us to overlook the true nature of this yeast product as a potent psychoactive drug.

Supernatural mediation in the experience of inebriation was imagined throughout the ancient world and alcoholic gods and goddesses flourished, including Dionysus and his Roman manifestation Bacchus, the Aztec god Tezcatzontecati, and the Sumerian goddess Ninkasi, mentioned earlier. Wine is the essential beverage of the Bible of course. Benjamin Franklin wrote, "Behold the rain which descends from heaven upon our vineyards, there it enters the roots of the vines, to be changed into wine, a constant proof that God loves us, and loves to see us happy." For those of agnostic persuasion, we may toast the evolutionary roots of our little fungus.

To appreciate the power of yeast we must define what it produces. Alcohol is one thing and many things. It refers to ethyl alcohol, or ethanol, and to members of the class of chemicals with a similar structure. Other alcohols include methanol (wood alcohol), sorbitol (a common sugar substitute), and menthol (from mint). Chemicals ending in -ol, which signifies the presence of an hydroxyl group (–OH), are alcohols. Here, however, alcohol will be used in the colloquial sense to refer to ethanol. This small molecule comprises a pair of carbon atoms bonded to hydrogen atoms and to a single hydroxyl group: $CH_3–CH_2–OH$.

Alcohol is a rare molecule in nature. Besides its formation by yeast, alcohol synthesis is limited to germinating seeds and a few kinds of bacteria. The bacteria tend to generate unpleasant flavors that can spoil beer and cider, so they offer little competition with yeast for the affections of brewers. Alcohol is also forged in the absence of biology in interstellar clouds. The largest molecular cloud close to the center of the Milky Way, called Sagittarius B2, contains enough alcohol for 10^{28} bottles of vodka, which, incidentally, would weigh five times as much as all of the planets in the Solar System.[7]

Yeast uses glucose and other sugars to fuel the dynamism of its cells. Energy is harvested from these molecules by splitting them into smaller parts and stripping energized electrons from their component atoms. Stripping electrons is called oxidation. When there is enough oxygen around, yeast has the option of breaking down glucose through two sets of reactions, capturing energy along the way, and leaving nothing in its wake but water and carbon dioxide (CO_2) (Figure 3). Stage 1 is called glycolysis and stage 2 is the citric acid cycle. This process of aerobic respiration is like a controlled burn that wrings the maximum amount of energy from the available fuel. Oxygen puts yeast into overdrive, and the fungus roars along like a Ferrari.

But yeast cells in beer wort and grape juice soon deplete the oxygen because the dissolved gas diffuses slowly through these sugary fluids. This blocks the more lucrative reactions of the citric acid cycle. Oxygen starvation can also limit the performance of a sports car and this is fixed with turbochargers that force more air into the fuel mixture. In similar fashion, brewers aerate their beer wort to optimize growth of the yeast, especially at the beginning of the fermentation process. This support is helpful in brewing,

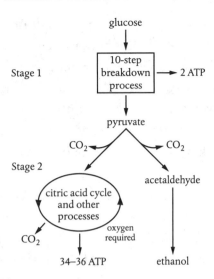

Fig. 3. Diagram of sugar metabolism in yeast. The initial ten-step breakdown of each molecule of glucose ($C_6H_{12}O_6$) to produce two molecules of pyruvate ($C_3H_4O_3$) is called glycolysis. The cell enjoys a net gain of two molecules of adenosine triphosphate (ATP) from glycolysis. ATP is a portable form of chemical energy that powers biochemical reactions in the cell. Subsequent metabolism follows two alternative paths. In aerobic respiration, the energy in the pyruvate molecules is captured via the citric acid cycle and other processes, with the release of CO_2 and the formation of additional molecules of ATP. In alcoholic fermentation, CO_2 is released with the formation of acetaldehyde, and ethanol is generated in a second reaction. Much of the energy present in the original glucose molecules remains in the ethanol released by fermenting cells.

but the fungus is proficient at adjusting to the loss of oxygen and can keep running in a different fashion. It does so by adopting an anaerobic burn or fermentation. This does not get as much energy out of the sugar molecules as aerobic respiration, but succeeds in meeting the immediate needs of a growing population of yeast cells. The residual energy is exhausted by the cell in the form of alcohol.

The compromise involved in leaving all those calories behind in the alcohol is well worth it because the chemical "exhaust" poisons every other fungus and bacterium that would like to compete for the sugars that allow yeast to thrive. This is so effective that yeast doubles down on its metabolic wager by choosing to produce alcohol even when there is plenty of oxygen around. Making do with this leaner diet is worth it if it keeps hordes of hungry microbes at bay. This strategy, called the Crabtree effect, is a critical part of the natural behavior of yeast.[8] It works because yeast has evolved an extraordinary tolerance to ethanol, and explains how the sugar fungus enjoys almost free rein to bubble away in rotting fruit and sweet plant sap. In brewing, the Crabtree effect allows yeast to make alcohol before and after oxygen levels fall.

Yeast produces alcohol until the level rises to between 10 and 15 percent, which kills the fungus. This process limits the alcohol content of beer and wine. Things work a little differently in nature, where yeast has the potential to keep growing after a period of fermentation by devouring its own alcohol. This metabolic dexterity allows the fungus to create, concentrate, and consume alcohol.

Yeast fell into its intimate relationship with alcohol after an abnormal act of copulation. Around 100 million years ago, when pterosaurs wheeled in the skies, a couple of yeast cells that had been budding in tree sap nudged against each other and mated. This liaison caused a genetic eruption called whole genome duplication.[9] Offspring produced from normal mating reactions have the same number of genes as either of their parents. In this instance, infant sugar fungi were born with double the number of genes: 10,000 instructions coded in the DNA of each cell rather than the usual 5,000-gene recipe for making yeast. As time passed, most of

8

the new copies of genes were removed from the genome and only 10 percent or so of the proteins made by today's yeast are encoded in genes derived from the genome duplication event.[10] But the consequences of this genetic eruption were momentous for us because yeast acquired the ability to process lots of glucose and make lots of alcohol.[11] This prepped the fungus to become our companion in brewing.

The timing of the genome duplication is critical to the story of yeast because the same doubling of information content happened to the Cretaceous ancestors of the flowering plants that went on to produce fleshy fruits.[12] The intersection of these events was crucial, because fleshy fruits are the natural source of the sugars that yeast uses to make alcohol. Yeast and the plants that manufacture its sweet food emerged from these ancient genetic explosions at around the same time.

Copying of individual genes is a major player in evolution because it provides opportunities for the replicas of genes to assume novel functions. Wholesale changes in the number of genes in an organism are more likely to be catastrophic unless the expression of all of these new instructions can be tightly regulated. Genome duplications may happen quite often on a timescale of tens of millions of years, but go unnoticed because most offspring delivered with more than the usual amount of DNA wither at birth. The profitable duplications of the genomes of yeast and flowering plants were extraordinary events in evolutionary history.

The discovery of yeast's effectiveness in raising bread dough was more serendipitous than its adoption for brewing, and would have escaped our attention if we had not made beer and wine first. Bread will never rise without a big shot of fresh yeast, and the most likely source was an accidental splash of aromatic froth from the top of a beer vat. Finding itself in moistened cereal flour, the yeast began with the two-stage process of aerobic respiration, digesting sugars in the dough as it was kneaded, releasing water and CO_2. Alcoholic fermentation would have kicked in too, as oxygen levels dropped, releasing alcohol and more bubbles of CO_2 into the dough.

The sight of the rising dough must have seemed magical to the early bakers. It remains so today. Bakers faithful to unleavened bread must have discarded these deformities until someone with a sense of adventure decided to put them in an oven. This was nothing less than an act of culinary heroism, on a par with the invention of ice cream.[13] The popularity of leavened bread across the centuries suggests that one of these pioneers escaped death by stoning and became a celebrity chef.

The fungal catalyst for brewing and baking is a marvelous micromachine crammed with fifty million proteins and other biomolecules. It has a diameter of 0.004 millimeters (4 μm), which is four times wider than a bacterium and half the diameter of a red blood cell.[14] When food is abundant, the ellipsoidal yeast cell swells for an hour or two, then squeezes a bud from its surface—effectively it gives birth. This yeastbirth culminates with the formation of a partition between mother and infant. Each bud bears a birth scar on the baby's surface, a solitary umbilicus, and leaves a matching pockmark, or birth scar, on her mother, who immediately begins to make a new bud at the opposite end of the cell.[15] The mother keeps budding, switching from one pole to the other. At times when separation is delayed, mothers carry the next generation around on their surface. Granddaughters may cling to daughters too, so that a multigenerational colony shaped like a ginger root casts itself in the bread dough or jiggles in beer froth.

After a few days, both ends of the mother cell are embellished with necklaces of birth scars—emblems of her productive life. Death comes for the yeast after a brief post-reproductive retirement. Meanwhile, her offspring keep growing and budding. The beer bubbles rise, and so too does the dough.

We use feminine nouns to describe budding yeast. She, the mother, always gives birth to daughter cells. This sounds right. "The father produces multiple daughter cells or sons on his surface," does not. After all, mothers give birth. This does not mean that yeast cells belong to a single gender, however, because the cells come in two versions, or mating types, designated **a** and a. These look the same and both types form buds. Their difference lies in their fragrance: the two mating types release distinctive kinds of chemical attractant when mating occurs. There is no obvious male yeast, just two kinds of parent in these sexual mergers.

When yeasts detect the perfume of the opposite mating type, their surfaces bulge so that both cells resemble gourds. The bulges are called "shmoos" in continuing deference to a rare display of professional merriment by geneticists, who related them to the promiscuous cartoon creatures introduced by the American artist Al Capp (who gave Americans the comic strip Li'l Abner) in 1948.[16] Shmoo-to-shmoo kissing creates fusion, like the fertilization of an animal egg by a sperm. The big cells created by fusion can keep growing, but when the supply of sugars dwindles they construct survival capsules called ascospores (Figure 4). Later, when more food becomes available, the spores germinate to release the next generation of yeast cells. This cycle of life has kept yeast going in something very close to its present form for the last five to ten million years.[17]

Besides its efforts in making wine, beer, and bread, yeast has become indispensable in the modern science of molecular genetics

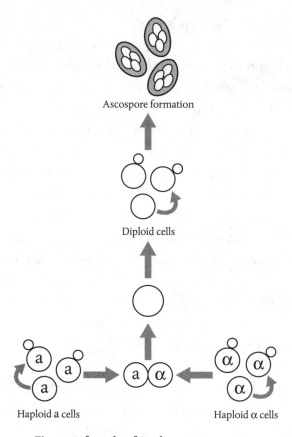

Fig. 4. Life cycle of *Saccharomyces cerevisiae*.

and as the workhorse of the biofuel industry. Yeast is an archetypal model system for research, chosen for its potential to uncover the ground rules of biology. The gut bacterium *Escherichia coli* was the first model microbe chosen by scientists who wanted to disentangle the molecular basis of heredity. Yeast is more complicated than this bacterium, but it has the advantage of telling us a lot more about human biology because it is much closer to us in its way of life. Bacterial genes are encoded in a single chromosome that is organized as a ring of DNA that is immersed in the fluid interior of the cell. Like our cells, yeast has multiple chromosomes housed within a separate cell compartment called the nucleus. Cells without a nucleus are called prokaryotic and cells with them are eukaryotic. Bacteria are prokaryotes; yeasts are eukaryotes, and so are humans. This means that what we learn about the way that yeast cells manage their lives may tell us a lot about how we get by.

Yeast has several attractions as an experimental organism besides its eukaryotic architecture. The fungus goes about its daily business with a single set of sixteen chromosomes. Cells like this, which contain a single set of chromosomes, are described as haploid. Animal and plant cells with two sets of chromosomes are diploid. Being diploid means that the effects of one version of a gene, called an allele, can be masked by another version of the same gene on the matching chromosome. This allows us, like Gregor Mendel's pea plants, to pass mutated versions of genes to our offspring without displaying their effects ourselves. An imbalance in fat metabolism, for example, can be passed to a child from both parents that carry a defective gene yet enjoy healthy levels of fat storage themselves. The errant version of the gene carried by both parents hides behind the uncorrupted gene on their second version of the same chromosome. But without a second copy of each chromosome in yeast, the effects of every bad gene are made immediately apparent. Mutation of a single gene affecting lipid management, for example, causes all the yeast cells to fill with globules of fat.[18] This vulnerability makes yeast a useful model for research on obesity and other diseases with a genetic basis.

Breakthroughs in the study of yeast genetics have driven the industrial uses of the fungus way beyond its innate talents as a brewer and baker. Genetically modified (GM) yeast strains are used to produce a range of drugs including human insulin, vaccines, and an injectable medicine to treat eye degeneration. It is easy to take the improvements in healthcare made possible by genetic engineering for granted, and tempting, given their astonishing profits, to malign the companies that make our medicines. This criticism, which can be justifiable, should be balanced with the reflection that insulin produced by yeast and bacteria allows millions of people with diabetes to control their blood sugar levels and circumvent limb amputation, blindness, and heart attacks. Life without this remarkable medicine used to be nasty, brutish, and shortened. The vaccine against human papillomavirus (HPV), which is the dominant cause of cervical cancer, is another yeast product whose power should disgrace a faith healer and stupefy a shaman.

Interesting yeast strains isolated from nature have also proven very useful without any genetic modification in the lab and are making biofuels from slurries of sugars extracted from corn and sugar cane. The United States specializes in corn (or maize) ethanol and Brazil is the leader in biofuel production using sugarcane. Corn kernels are packed with starch that has to be converted into sugars before yeast can make any ethanol. This chemical transformation is achieved by cooking a mash of kernels with powerful enzymes that come from other fungi and from bacteria. Sugarcane is a superior crop plant for bioethanol, because juice squeezed from its fibrous stems are already rich in sugars that yeast will convert into ethanol. Ethanol from corn and sugarcane are examples of first-generation biofuels.

Second-generation biofuels come from fibrous agricultural waste, often called biomass. This represents a massive source of untapped energy. Most of the substance of crop residues like wheat straw is made from sugars that could be converted into ethanol, but they are locked inside larger molecules called poly-saccharides that resist digestion by yeast. If we could create a single yeast strain, or a community of yeasts, which could release the sugars from this biomass we could solve our energy needs long into the future and abandon fossil fuels for good. The Holy Grail for biotechnology is a single yeast strain, genetically modified to make ethanol in this fashion, but we are a long way from transforming the sugar fungus into a microbe that can thrive on fiber. As the planet warms, the quest for this ultimate biofuel factory may become recognized as the most important mission in the history of science.

The story of yeast is more than a biography of a single species. Yeast is so many different things. The name "yeast" refers to *the* yeast, *Saccharomyces cerevisiae*, the vaginal yeast *Candida albicans*, and many other species of single-celled fungi. Whether or not they ferment sugars and release ethanol, yeasts live on a diet of simple foods. In nature they contribute to the decomposition part of the carbon cycle by digesting substances produced by plants and animals or dissolving the post-mortem remains of their tissues. All of these inconspicuous fungi share an affinity for growth in fluid habitats, and many of them are able to cope with diminishing levels of oxygen that asphyxiate other forms of life.

The fifth edition of the yeastologists' bible, *The Yeasts: A Taxonomic Study*, describes nearly 1,500 species.[19] The majority of them, including *Saccharomyces* and *Candida*, are ascomycete fungi, members of the Phylum Ascomycota. Yeasts are the least comely members of this brand of organisms whose larger representatives include brightly colored rock lichens, woodland cups decorated with eyelashes, and fragrant morels that plead for cream sauce.[20] Harvard mycologist Roland Thaxter was so captivated by the anatomy of the ornamented flasks of minuscule species of ascomycete that grow on insect cuticles that he spent thirty years illustrating their charms (Figure 5).[21] *Saccharomyces* has never had an artistic champion of Thaxter's talents. If one were inspired, he or she would be classed as a minimalist.

The significance of *Saccharomyces* in human affairs, and as a model for cell biological and genetic research, tends to overshadow the value of other yeasts, but one particularly important species has played a huge role in scientific progress. Sir Paul Nurse, Sir Timothy Hunt, and the unknighted American scientist Leland Hartwell, were awarded the 2001 Nobel Prize in Physiology and Medicine for their work on the mechanisms that govern the division of cells.[22] The model for the experiments by Paul Nurse was the fission yeast, *Schizosaccharomyces pombe*, rather than the sugar fungus.

When we consider the lifestyles of other yeasts, the picture of mankind's helpmate becomes more convoluted with the realization that our bodies provide many of these microscopic fungi with their living. Our skin surface, ears, noses, mouths, vaginas, and digestive systems teem with a variety of yeasts. The activities of many of these species are unclear, and the only reason we know about them at all is that their genes show up on cotton swabs and in fecal samples. The gut yeasts probably work alongside the more numerous bacteria in the digestion of our food. Yeasts on our skin also interact with bacteria and are considered part of the healthy microbiome—the galaxy of microorganisms that we carry on and inside our bodies. Dandruff is related to the growth of a yeast called *Malassezia globosa* in skin flakes, and it is controlled by antifungal agents incorporated into medicated shampoos. *Malassezia* consumes fatty sebum secreted into hair follicles and competes for food with resident *Demodex* mites. Swarming with yeast and mites, we are walking ecosystems.

Fig. 5. Roland Thaxter's illustration of flasks, or thalli, of species of Laboulbeniales (ascomycete fungi) that grow on the exoskeleton of beetles and other insects. The largest thallus in this group is 0.8 millimeters long.

Candida albicans is a more important yeast from a clinical perspective. It is a natural component of the microbiome that becomes a problem when the healthy populations of microorganisms in the vagina are disrupted by antibiotic treatment. *Candida* can also be very dangerous when it invades the tissues of patients with damaged immune systems. Other yeasts can also cause these illnesses, called opportunistic infections, including black yeasts with heavily pigmented cell walls. The phenomenon of the opportunistic infection confuses the ostensible distinction between a fungus that promotes our well-being and a fungus that makes us ill. Even the sugar fungus can turn nasty if it is implanted in someone during surgery.

Putting aside a handful of case histories in the medical literature describing human infections involving the sugar fungus, the blessings of our favorite yeast are everywhere when you look for them. Nicholson's Pub in Cincinnati has served as a sanctuary for me for many years. The fruits of our association with yeast are manifest in this saloon, with beers on draft, bottles of malt whiskey glinting on glass shelves backed by mirrors, and hot rolls in bread baskets on tables. Leaving the pub after a recent visit, merging with the stream of people on the sidewalk, I noticed that more than half of the pedestrians were marked with a gray cross on their forehead. It was Ash Wednesday. Cincinnati Catholics decorated by the priests at Holy Communion had also received the blessings of bread and wine, the body and blood of Christ. The sacramental wafers were unleavened, but every sip of the wine was fermented by the fungus. And the cars and trucks streaming through the city streets that afternoon were running on fuel blended with bioethanol made by *Saccharomyces*. Civilization runs on a deep symbiosis with the sugar fungus.

2

~

Yeast of Eden

Drink

At night in the rainforests of Malaysia, drops of nectar collected within the huge flower stalks of the bertam palm are fermented by a community of yeasts. The resulting plume of alcohol that wafts through the woods makes pen-tailed treeshrews twitch their nostrils in anticipation (Figure 6).[1] Climbing the palms after nightfall, these little mammals with feathery tails guzzle enough of the sweet liquor to match the alcohol consumption of a human who enjoys a strong cocktail followed by an entire bottle of wine.[2] Chemical analysis of hair samples from the treeshrews shows that they are chronic drinkers, though videos of them recorded with night-vision cameras reveal no signs of drunken or disorderly behavior. They move around the trees with the steadiness of tightrope walkers, sidestepping the fearsome spines that ornament the leaves and flowers. Their evident sobriety is remarkable. Few of us would fare well if we tried to negotiate a fence of razor wire after a bottle of claret.

There is no entertainment value in alcohol for a treeshrew. Inebriation would be a death sentence for an animal navigating a forest of spiny plants and sharp-toothed predators. Alcohol is merely a signal for the treeshrew that sugary nectar is nearby, the reward for its unconscious transfer of pollen from tree to tree.

Fig. 6. Pen-tailed treeshrew, *Ptilocercus lowii*.

Researchers speculate that the mammal bypasses any side effects by burning off the ethanol very quickly, detoxifying the nectar and enjoying this valuable source of natural sweetness.

Yeasts grow wherever they find sugars, and the tropical palms offer an attractive nursery for billions of budding cells. Bertam palms produce nectar for several weeks, exuding syrup between the petals of their woody flowers. Fermentation takes place in special brewing chambers where the fluid is trapped between the tightly packed flower buds, and becomes so feverish that the nectar froths with bubbles of CO_2. Because alcohol evaporates, the strength of the brew drops quite quickly, but the continuous production of the sugary liquor by the palm keeps the yeasts fed and the alcohol flowing, maintaining the bait for pollinators from the surrounding forest. This intentional explanation of the rainforest brewery—intentional, meaning *adaptive* on the part of the palm— links the anatomy of the flower chambers to the fermentation of nectar, and the role of alcohol as an advertisement to pollinators. The pen-tailed treeshrew is not the only visitor. The Sunda slow loris is another nocturnal drinker, and rats and squirrels explore the aromatic flowers day and night (Figure 7).

We do not know how often brewing takes place in flowers of other plant species as a pollination aid. It seems logical that plants which produce their own scents and use colorful petals to attract insects would have little need for yeast and its alcohol.[3] Nevertheless, a variety of yeasts are found in the flowers of many plant species,

Fig. 7. Sunda slow loris, *Nycticebus coucang*.

and a number of these fungi are capable of fermenting sugars and producing a plethora of odors. The evolution of an unusual plant called Solomon's lily, which grows in Israel, provides one line of evidence that yeasts may be more important for attracting pollinators than botanists have recognized. Close relatives of this flower attract beetles and other insect pollinators by smelling of dung and urine. Solomon's lily takes a different tack entirely, releasing a bouquet that resembles a fruity wine.[4] It performs this trick on its own, without secreting nectar and cultivating any yeasts. Fruit flies attracted by the odor of yeasts in rotting fruit are fooled into carrying pollen without enjoying any nectar for their services. Solomon's lily is unlikely to be the only plant to have hit upon this ruse.

Overripe or rotting fruit is the most reliable source of alcohol in the wild, and there have been lots of claims about animal inebriation associated with fruit consumption or frugivory. Few of these appear to be true. Fruit bats seem a good test case for animal drunkenness because their foods are likely to be fermented in nature. To assess the alcohol tolerance of wild fruit bats in Belize, Canadian investigators captured the animals in mist nets, fed them with sugar water spiked with alcohol, and evaluated their flying skills in an enclosed obstacle course.[5] The fruit bats passed this sobriety test, dodging plastic chains hanging in their flight path without any collisions. They were as graceful in the air as treeshrews on their palms.

Egyptian fruit bats did not fare as well in another study. The faces of these lovely animals resemble dogs. They have soft fur, "wings that feel like pantyhose," as one chiropterologist put it, and wild dates are their favorite food. When these bats were fed alcohol, they flapped their wings with less enthusiasm than usual as they flew from end to end in their enclosures. More telling was the confusion in their echolocation calls that researchers described as "analogous to speech impairment in inebriated humans."[6] The difference in behavior between bat species may lie in the higher levels of alcohol found in the fleshy fruits of Neotropical forests compared with the dates guzzled by the Mediterranean bats. Like the treeshrews, there is a premium upon alcohol detoxification for the bats from Central America, whereas the Egyptian bats have not had any need to evolve mechanisms to neutralize this foreign dietary danger.

Elephants seem like a much better bet as natural boozers. Traveling in Natal in the 1830s, Louis-Adulphe Delegorgue, a French naturalist who expressed a sickening glee in his remembrance of slaughtering herds of elephants, reported, "The elephant has in common with man a predilection for a gentle warming of the brain induced by fruit which have been fermented by the action of the sun."[7] He was commenting on the supposed effects of the fruit of the marula tree. This anecdote remains part of local folklore and is encouraged by online videos of elephants and other animals falling over after gorging themselves on the succulent drupes—the botanical name for stone fruits like mangoes and olives—of this African tree. Despite this footage, calculations show that a big mammal could never eat enough fruit to get drunk.[8] If the pachyderms do act strangely after eating rotting marula fruit, it is likely that they are stupefied by something other than ethanol—the burst of sugar perhaps, or an unknown hallucinogen. Nevertheless, the popular association of marula trees with elephants is reproduced in the tusked bull on the label of Amarula Cream Liqueur, which is fermented from the fruit in South Africa. The distillation process pushes the alcohol content of this sweet drink to 17 percent, which is much stronger than the proof of the fruit fermenting beneath the trees.

The attraction of elephants to fermented drinks is illustrated by the behavior of the first live specimen exhibited in America in the eighteenth century. This "most respectable animal" was said to enjoy "all kinds of spirituous liquors," and drank as many as thirty bottles of porter (dark beer) in a day, "drawing the corks with his trunk."[9] Perhaps he recognized the scent of alcohol plumes from the savannah of his childhood. Alcohol appeals to other animals too. Chimpanzees in Guinea steal fermenting palm wine from plastic containers placed by tappers in raffia trees.[10] The apes climb into the trees and scoop out the fermenting wine using folded or crumpled leaves. After a drinking session, the chimps show little evidence of intoxication. They appear to enjoy the sweetness of the wine without consuming enough to make them unsteady.

The rarity of animal intoxication in nature does not mean that humans are especially vulnerable to the effects of alcohol. Without the ingenuity of brewers and vintners, whiskey distillers, and palm wine wizards we would be as sober as the rest of the animal kingdom. Access is the key. This is evident from the way that poachers are able to catch game birds by feeding them grains soaked in alcohol and laboratory rodents can be schooled as alcoholics for addiction research.

Some years before the 2008 exposé on the pen-tailed tree-shrew's attraction to alcohol, biologist Robert Dudley proposed the "drunken monkey hypothesis" to explain the origins of human alcoholism.[11] This was a prescient idea that drew on information from multiple fields of research. It was an interdisciplinary theory par excellence. Dudley suggested that our attraction to alcohol is an extension of behavior that evolved among earlier primates and enabled them to find ripe and overripe fruit.

The "drunken monkey" essence of humanity is written in our genes and expressed in our metabolism. Human tissues are primed to deal with alcohol: our bodies expect to encounter it. We process the molecule via a pair of biochemical reactions that reorganize its structure in preparation for breaking it apart to harvest energy. Biochemical reactions are sped-up, or catalyzed, by proteins called enzymes. Alcohol dehydrogenases are enzymes that carry out the first reaction, which produces acetaldehyde, and acetaldehyde dehydrogenases convert acetaldehyde into acetic acid. Acetic acid is consumed in the furnaces in our cells called mitochondria, and the chemical energy from these reactions is exported in the form of the ATP molecules that fuel the life of the cell. Different versions of these dehydrogenase enzymes affect the way that we handle alcohol and their distribution varies across human populations. People who inherit genes that encode very efficient types of alcohol dehydrogenase are able to metabolize alcohol more swiftly than individuals who express less vigorous versions of the enzyme. Effective alcohol metabolism can be perilous for some partygoers, however, because it encourages the celebrant to imbibe more, over a longer period of time, and descend, sooner or later, into the depths of complete intoxication. This may explain why I discovered myself one morning in 1986 on a train heading to New England, with no memory of the previous twenty-four hours beyond a blurred recollection of evening festivities in Manhattan.

The acetaldehyde resulting from the second reaction managed by these enzymes is responsible for some of the symptoms of a hangover, which means that acetaldehyde breakdown is critical for anyone who enjoys drinking. Thirty percent of Asians of Han Chinese, Taiwanese, and Japanese ethnicity inherit an aldehyde dehydrogenase that is almost useless in this regard.[12] Many people with this genetic background tend to avoid alcohol consumption, but some put up with the skin blotching known as the alcohol flush reaction, and other symptoms of aldehyde accumulation, and drink as much as Europeans. This determination, coupled with wide variations in the genetics controlling aldehyde breakdown, explains why the Asian market for malt whiskey and other spirits is so buoyant.

The genetics of alcohol metabolism in humans are further complicated by the production of different versions of the crucial enzymes in different parts of the body. Alcohol dehydrogenases are encoded by seven genes that are organized in a cluster on human chromosome 4. One of these enzymes, abbreviated as ADH4, acts on alcohol on the tongue as we hold champagne in our mouths for a few seconds, in the esophagus when we swallow, and in the stomach where the sparkling wine is introduced to, say, raw oysters. ADH4 is hyperefficient at removing hydrogen atoms from yeast ethanol to form acetaldehyde: $C_2H_6O \rightarrow C_2H_4O$. The reason that it does this so well lies in a mutation in the *ADH4* gene that occurred ten million years ago. The ADH4 enzyme is a protein made from a coiled and folded chain of 380 amino acids. The mutation replaced one of these amino acids in the chain, called alanine, with a larger amino acid called valine, and this made the enzyme work a whole lot better. Mutations of this kind are called *substitutions*. We know this happened early in the evolution of the great apes, because exactly the same mutation is present in the *ADH4* gene of gorillas, chimpanzees, and bonobos.[13]

Beyond the apes, in the family tree of primates that has been branching since the disappearance of the dinosaurs, the mutation shows up in just one other animal: the aye-aye. Aye-ayes are the endangered Madagascan lemurs with long fingers used for the percussive tapping of wood and extraction of the fat larvae of long-horn beetles. A possible explanation for the presence and persistence of the mutation in aye-ayes lies in the observation that the lemurs pollinate a native palm tree. Like the treeshrews, aye-ayes may be instinctive tipplers.

When biologists look at evolutionary trees like this, which show a spotty distribution of a particular characteristic, they consider two possible explanations. The first is that an ancestral primate possessed this version of the ADH4 enzyme and that it disappeared in some of its descendants and was retained in others. This phenomenon, called *secondary loss,* does not seem to work for the super-enzyme because it is just so rare. *Convergent evolution* is the alternative model for the appearance of shared characteristics, and explains the evolution of the alcohol dehydrogenase gene: the simple mutation in the alcohol dehydrogenase gene occurred at least twice and its value allowed it to spread independently among aye-ayes and our ancestors.

The spread of the great ape version of the alcohol dehydrogenase enzyme and its conservation to this day is probably explained by diet. Apes evolved in rainforests where an abundance of fallen fruit advertised by alcohol plumes offered a superb source of calories. Gorillas, chimpanzees, bonobos, and humans learned to gather these delicacies by walking on the ground (with or without relying on knuckles for stability). When fruit became an important part of the ape diet, a premium was placed on the most efficient machinery for removing alcohol from the body. It is interesting that the ability to process alcohol did not evolve among the hundreds of other frugivorous primates that must consume fermenting fruit from time to time.

Humans are without peer as drunken monkeys, because we are the only animals who figured out how to work with yeast to brew enough alcohol to intoxicate ourselves. In *Paradise Lost*, John Milton assures us that Eve served Archangel Raphael unfermented grape juice, "inoffensive moust [must]," when he visited the first couple in Eden with warnings about Satan. Our genes say different, and suggest that we have been enjoying highly *offensive* must from the beginning. Deliberate fermentation may actually have been invented long before humans were trotting around in Eden or anywhere else. Here is how it (probably) happened. One fine day, some early species of hominid drank some palm sap that had been drained into a calabash or upturned turtle shell and left out in the sun for too long. News of the resulting feeling of euphoria after he or she glugged the spoiled brew was shared among these big-brained animals and the technology of fermentation was born.

If palm wine was the earliest fermented drink, palm wine drunkards were the first alcoholics. The discovery of 3.3 million-year-old stone tools in Kenya widens the plausible timeframe for the first experiments with fermentation.[14] The beauty of this palm wine hypothesis lies in the automatic fashion with which the mere collection of sugary palm sap leads to its transformation into alcohol. Stone Age brewing was fated. Microscopic starch grains found on 105,000-year-old stone tools in Africa provide some clues about the first drinks brewed by our species.[15] Pear-shaped starch grains match granules from the African wine palm, *Hyphaene petersiana*. This plant is tapped to produce "Ombike," which is a traditional home-brewed liquor of the Aawambo people in Namibia and Angola. Starch grains from wild sorghum seeds were also found on stone tools. Sorghum beer has been brewed for millennia and remains very popular in southern Africa.

Brewing palm wine is so simple once the raw ingredients are sourced that Ombike, or something very similar, is likely to have been the first drink that was fermented by design. Getting the sap is a dangerous task for practitioners called tappers or tapsters, who climb into the crown of the tree, cut the flower stalks, and drain the sap into pots. A second approach is to remove the branches at the top of the palm and cut the terminal bud, which kills the tree. Either way, climbing palm trees requires skill. This much is clear from Amos Tutuola's bestselling tale, *The Palm-Wine Drinkard*, published in 1952.[16] The narrator, a Nigerian Orpheus addicted to palm wine, is devastated when his personal tapster falls from a tree and dies. Unable to find a replacement employee for this skilled work, he sets out to resurrect the tapster from the underworld. A third and much safer method of tapping is to fell the tree and bleed it on the ground. Palm sap collected in a pot begins fermenting as soon as it is exposed to the air. When the pots are left in the sun for a couple of hours, wild yeast produces a sweet aromatic liquor with an alcohol content comparable to beer.

Without the deliberate addition of yeasts to the palm sap, the brew is fermented by the local strains that find their way into the pots. Analysis of palm wine in Cameroon revealed a mixture of yeasts in the earliest phases of the fermentation.[17] This ragbag of wild fungi jostles for control of the sugar, but is soon replaced by a single strain of *Saccharomyces* that clobbers the competition by soaking them with alcohol. The same thing happens wherever palm sap is exposed to the air. The complex mixture of species that develops at the beginning of fermentation is always overthrown by a monoculture of the sugar fungus. Palm wines are gardens of live yeast and have no shelf life at all. They are often consumed after a few hours of brewing, but become stronger if the fermentation is allowed to continue. In these calabash-aged wines, the sweetness of the early fermentation gives way to sourness and acidity that is preferred by some connoisseurs. Even longer fermentations encourage bacterial growth that turns wine into vinegar. Before this desecration takes place, younger wines can be distilled to bump up the alcohol concentration to more intoxicating levels. Palm wine distilleries are quite common in Africa and Asia.

The earliest clear archaeological evidence for brewing comes from chemical analysis of 8,200- to 8,600-year-old pottery shards from China, which indicates that Neolithic villagers fermented a drink from rice, honey, and fruit.[18] The ingredients suggest that this would have tasted like modern rice wine. A cuneiform tablet from southern Iraq details the 5,000-year-old beer ration for Mesopotamian workers using the impression of an upright jar with a pointed base as the symbol for beer. Ceramic jars of similar age from Egypt, recovered from the tomb of Pharaoh Scorpion I, have a yellowish deposit on the inside from which investigators claim to have identified yeast DNA. Grape pips found at the same site in Abydos in Upper Egypt support the conclusion that the residue in the jars came from yeast that was fermenting grape juice.

Ancient pots and tablets evoke the trappings of "sedentism"— the term that cultural anthropologists use for living in one place rather than as nomads, and demonstrate that brewing techniques were supported by early agriculture. According to this microbiological interpretation of history, yeast qualifies as the *primum mobile*, or author of the modern world. The claim that civilization was kindled by our love of alcohol hinges on the proposition that cereal agriculture and the attendant human settlement were designed to provide the raw materials for brewers. This idea was first voiced in the 1950s and is the counterpoint to the "baking came first" view of civilization. As long as human populations remained quite small, hunting and gathering met their nutritional needs. There was no impetus for sedentism. Brewing presented greater difficulties for nomads. To go beyond palm wine, or the occasional drop of beer made from wild sorghum, the cultivation of grasses and grapes was needed to ensure a reliable supply of alcohol. Some posit this as the reason that civilization began in villages surrounded by golden fields of barley and rows of grapevines on the hills.

The eminent anthropologist, Claude Lévi-Strauss, regarded the invention of brewing as a symbol for the passage of humans from "nature to culture."[19] Lévi-Strauss was interested in the idea that mead was the original drink that encouraged settlement rather than beer or wine, but the microbiological agent was the same. The fungus ferments the sugar in honey just as effectively as it makes alcohol from cereal grains and grape berries. Lévi-Strauss meant culture in the sense of human behavior, but, with a twist, the phrase nature to culture applies to *Saccharomyces* too—since yeast was taken from nature and cultured by brewers.

We have seen already that the human genome is adapted in varying degrees to deal with alcohol. Because every drop of alcohol comes from yeast, there is a sense in which the genetics and practice of human drinking were largely crafted by the sugar fungus. If agriculture and civilization were founded on the needs of brewers, it follows that we have been tamed by *Saccharomyces*. This is particularly compelling for those of us who hunt for the corkscrew at the end of a tiring day. There is a certain truth to this idea of an ancient co-dependence. Yeast, as the silent partner in this venture, also shows signs of domestication in its genome. For a long time it was believed that *Saccharomyces* was a sort of house-trained pet, the microbial equivalent of the domestic cat. Different strains of wine and beer yeasts are like cat breeds, but microbiologists had no idea where they had come from until they began using molecular methods to search for strains of the fungus in the wild.

The presence of feral yeast close to domestic vineyards was no surprise, given that it is a microorganism that escaped from wine presses all the time, but, as the search continued, wild strains of *Saccharomyces* were found on oak trees—on their bark, leaves, acorns, and in the surrounding soil.[20] Yeast showed up in forests far from any vineyards. These strains had never been domesticated, and had never feasted in a wine barrel or a beer vat. Most of the diversity existed among fungi collected in China.[21] When we explore the genetic diversity of humans, we learn that variations on our bipedal theme are widest in Africa. This is the evidence, along with a lot of fossils from the Rift Valley, that we came from Africa. And just as *Homo sapiens* is a species of African ape, *Saccharomyces cerevisiae* is a species of Asian fungus.

Once out of China, it moved everywhere. The strains employed in winemaking seem to have migrated westward from the Fertile Crescent around 10,000 years ago.[22] Some botanists think that domesticated varieties of grapes have a more recent origin, beginning around 7,000 years ago. This mismatch is significant if we assume that grape cultivation and winemaking went hand in hand. It is possible, however, that early vintners pressed wild grapes wherever they found them and, unknowingly, carried yeast strains that were good fermenters on their pots and presses. The farther that these migrants penetrated into Europe, the more the yeasts changed through mutation, which explains why the European strains are so different from their Middle Eastern relations. The

10,000-year origin for wine yeasts is compelling, because the same date applies to the domestication of cats, many farm animals, and crop plants. Cats thrived in villages, where they caught the rodents that consumed our stored grain (or, according to another idea, simply stuck around, as cats are wont to do, without contributing anything tangible to the function of the community). Goats and sheep were tamed around the same time, and barley and wheat were bred from their wild relatives. Pigs were farmed a bit earlier in the Near East, chickens more recently.

The timeline for these historical events is based on a growing archive of archaeological finds and from molecular clocks made from DNA. The strings of sequences of As, Ts, Gs, and Cs, which make up everyone's genome—GTGCAATCAC and so on, for three billion letters in the glory that is us and twelve million in yeast—are altered by mutations as time passes. One letter gets switched to another, another letter is removed, and new code is inserted into an existing sequence. Many of these mutations do not make it into the next generation because the individuals in whom they occur do not pass them on to their offspring. Often, this failure to copy occurs because the mutation is harmful. But when mutations occur in noncoding, or neutral, regions of the genome, where the DNA sequences are not used to make anything, natural selection does not have the opportunity to remove them. The slow accumulation of these neutral mutations serves as the molecular clock, which can be read by comparing the DNA sequences between species or within strains of a single species. Sequence differences provide a rough estimate of how long these organisms have followed separate evolutionary paths.

The genetic distinction between yeast strains from wine and yeast strains isolated from soil is consistent with their pursuit of separate lives for 10,000 years.[23] Although beer brewing may have been perfected before winemaking, it is impossible to track the ancient movements of the beer yeasts because the oldest surviving breweries use yeast strains of relatively recent origin. Some lovely work has been done on the evolution of the yeasts used to make ales and lager, but their history plays out over mere centuries rather than millennia.[24] The transfer of yeasts from one batch of beer to the next, called "back-slopping," has always tended to segregate brewing yeasts from their relatives in nature. If the brewer is diligent about this, mating between the domesticated strains and wild yeasts is prevented. This encourages the evolution of characteristics that might render the fungus uncompetitive in nature but very useful for beer-making. Clear evidence of this artificial selection can be traced back to the seventeenth century among yeasts used in European and American breweries today.[25] This domestication process allowed the evolution of yeasts with extra copies of genes that enhanced their ability to feed on the mixtures of sugars released from the malting process. It is possible that some American strains are the direct descendants of yeasts imported by settlers from Jacobean England.

The historical movements of yeasts are inseparable from the patterns of migration by human brewers and winemakers. We have traveled bud in hand and covered the globe, in much the same way that the distribution of cereal crops and domesticated animals matches human movements. Yeast owes as much to us as vice versa because it does not get around very well on its own.

Saccharomyces is not a typical fungus. The spores of most fungi are dispersed by air currents. Some of the common molds like *Aspergillus* and *Penicillium*—fungi that bloom as colored blotches on dairy products, form spores on stalks, and shed them as dust in the wind. Other fungi use more active mechanisms to get airborne, including pressurized squirt guns, jumps powered by water drops, and raindrop splashes from tiny nests.[26] Hundreds or thousands of these spore movements occur in the blink of an eye, and high-speed video cameras are needed to slow the motion to see what happens. The combination of these passive and active mechanisms distributes millions of tons of spores into the air every year, affecting the chemistry of the atmosphere and causing hundreds of millions of asthmatics to wheeze.

Yeast does not migrate in any of these ways. Air filters viewed with a microscope are clogged with sparkling spores, but none of them come from the sugar fungus. For a long time, descriptions of traditional winemaking—without the deliberate addition of yeast by the vintner—referred to the inoculation of grape must by yeast cells stuck to the surface of grapes. More recently it was shown that *Saccharomyces* does not multiply on grapes as they ripen on vines, at least if they are undamaged. Bruised fruits with broken skins are more likely to harbor the fungus, but, even then, yeast is not found on more than one in four berries.[27] The idea that yeast settles on the grape must during pressing is more compelling, though the air inside wineries has very low levels of yeast until the grapes are crushed and the fermentation is in progress.[28]

The same uncertainty about the arrival of yeast applies to palm wine and other fermentations. Yeast will start growing the moment it finds itself in a pot of palm sap and tastes sugar. The fungus may come from the inside of the pot, the skin and clothing of the artisan, or the spoon he uses to stir the brew. After all, it only takes a single cell to get things started and exponential growth takes care of the rest. When yeast is growing at its swiftest rate, with the number of cells doubling every ninety minutes, a founding population of one hundred cells could grow to 430 billion in two days. These short-lived residues of yeast probably work very well when brewing is a frequent activity in the same location, but this does not explain how yeast makes its appearance in new places.

The answer to yeast's mobility lies in its presence in the guts of insects, particularly social wasps.[29] Free yeast cells may not be airborne but they fly as passengers inside wasps. With so much insect activity in vineyards and the powerful attraction of insects to ripened fruit, yeast is deposited by insects when they gorge on fallen berries. European hornets, which are types of wasps, build paper nests in spring that house hundreds of workers. They are predators of honeybees and other insects, which they feed to their larvae. As the year progresses, they turn to a sugary diet as fruit becomes plentiful. Hornets carry a community of yeasts inside them, transmitting the fungus as they buzz between fallen apples and pears, land on ripe grapes, feed and defecate, and attack anything that threatens their nests.

Winemaking is of course big business, and relying on the natural contamination of grape must via insects, or some other means, is considered too haphazard for most modern vintners. The use of a particular strain with a fully sequenced genome is viewed as essential for a mass-marketed wine. Italy is the world leader in the wine industry, with an annual production of five billion liters.[30] Although many Tuscan winemakers add particular starter strains to the grape must to ferment the glorious varietals of this region,[31] the natural yeasts carried by hornets and other wasps in the vineyards continue to be players in the surrounding ecosystem. Winemakers who trust their vintages to spontaneous fermentation rely on yeasts on the crushed grapes, and some of these strains are introduced by these insects. Queens carry yeast through the winter, transmitting the fungus from one nest as it shuts down to a new one the following spring. Yeast survival in insects was monitored in one experiment by feeding wasps with a genetically engineered strain that glows green in ultraviolet light, allowing the animals to overwinter, and then recovering fluorescent yeasts from their guts in the spring.[32] The thick-walled ascospores that form after yeast cells mate seem to help the fungus survive during its passage through the insect gut.[33]

So to get around, wild yeast needs insects—wasps, hornets, fruit flies. Not content with the scent of its alcohol plumes to attract them, yeast perfumes its odor with a cocktail of volatile organic compounds called acetate esters. In an elegant series of experiments, Belgian scientists created a mutant yeast that lacked two copies of a gene that controls the synthesis of these aromatic molecules.[34] (The presence of two copies of this DNA sequence takes us back to the genome duplication event described in Chapter 1.) When fruit flies that had been starved were placed in a miniature arena with colonies of normal and scentless mutant yeast, the insects raced toward the fragrant ones. The starving insects suffered further at the hands of the researchers who secured them with wax, exposed a window into their brains by removing a scrap of exoskeleton from their tiny heads, and monitored their neurological activity using a fluorescent dye. As the tethered flies were misted with yeast volatiles, bright green flashes of light spread across the lobes of their brains connected to their antennae. Other investigators stuck electrodes into the fly brains and measured the electrical impulses in the nerve cells as they were challenged with odors from *Saccharomyces* and those of other yeasts.[35] The neurologial responses of the flies showed that they could tell the difference.

The hard-wired attraction of fruit flies to the sugar fungus helps them to find food in the form of fermenting fruit. At the same time, yeast is furnished with a convenient form of transportation inside the hungry insects. The acetate esters produced by yeast that appeal to the flies are also part of the smell of wine and beer that we find so intoxicating. It seems that the oenophile who assesses a wine with an expert proboscis is responding to the same whiff of fungus as a fruit fly. We certainly compete for the same vintages. Within a few minutes, a glass of Cabernet left unattended in my garden is besieged by a congregation of these pests, most of which drown in the ruby liquid. Like treeshrews and humans, the flies possess the enzymes needed to oxidize ethanol. These evolved as a detoxification system for the insects. Although they can use ethanol as an energy source, getting rid of the stuff is more important from the viewpoint of survival.

Unlike treeshrews, however, fruit flies display all the signs of drunkenness after binging. When they are exposed to high concentrations of alcohol in the form of vapor they become excited, move around more swiftly than usual and bump into obstacles, fall over, and, finally, go to sleep.[36] Flies given a choice between sugar water with and without alcohol slurp the syrup containing up to 26 percent alcohol. The insects' preference for drinking alcohol increases over several days.[37] Their sexual behavior changes too. Males that become habitual drinkers lose their normal inhibitions and increase their courtship behavior toward other males as well as toward females.[38] Males that engage in successful mating when sober are less interested in drinking afterwards, and males whose sexual advances are spurned by females consume more alcohol in response to rejection.[39] Substitute *human* for *fly* in this paragraph and almost nothing changes. Fruit flies exposed to alcohol as larvae develop more slowly than normal, have smaller brains, and grow into smaller adults.[40] The similarities between flies and humans deepen with each study.

The reason that fruit flies and humans have comparable responses to alcohol lies in the resemblance of our nervous systems. The fact that the fly brain has 135,000 neurons compared with the eighty-six billion cells inside our skulls is irrelevant. Alcohol has the same effects on the underlying circuitry and both species seem to enjoy it, thirst for it, and run the risk of being destroyed by it. This predilection is not a problem for fruit flies unless they get into our drinks.

With humans, on the other hand, brewing and drinking are defining features of our species: *Homo sapiens, a bipedal ape that makes beer and wine wherever it settles and drinks alcohol for pleasure.* We have always done it, and we have the genes to prove it. There is no single alcoholism gene, but suites of genes whose expression has a powerful effect on one's attraction to drink. Environment is crucial too, with influences varying from the availability of alcohol to life experiences that drive some of us toward the bottle. The reasons the genes that confer a propensity toward excessive drinking have persisted for so long raises some troubling questions. There must be an advantage to heavy drinking that outweighs the disadvantages. That premium has to lie in sex. People who have enjoyed alcohol in previous generations must have been successful at reproduction. The details of this sociobiology are not known, but there are plenty of possibilities. One is that the activities of cereal and grape cultivation, brewing, and drinking are so stabilizing for a community that participants leave plenty of offspring. This is the "it takes a village of drinkers" model of human progress, or *Saccharomyces* the peacemaker.

Another explanation is more direct, which is that when young men and women drink they have a greater tendency to become parents. A simple lowering of social inhibitions is often sufficient explanation, but changes in aggression and compliance are at work too. The effects of alcohol consumption on social behavior are surprisingly similar to those of the hormone oxytocin. Oxytocin is secreted from the pituitary gland, and when the levels in the brain are increased by nasal spray, test subjects become less anxious, express greater empathy, and rate other people as more trustworthy and attractive.[41] The list of positive feelings provoked by the hormone is extensive. Oxytocin reduces stress and anxiety by stimulating the release of the inhibitory neurotransmitter gamma-aminobutyric acid, or GABA, which dampens the excitability of neurons throughout the nervous system. Alcohol does not stimulate GABA release in this way, but certainly has the same sort of mellowing influence on the nervous system. It also plays with our serotonin and dopamine levels, managing to work as a mood elevator, at least temporarily, and a depressant after a serious drinking bout.

Alcohol can, of course, produce aggressive behavior in some drinkers. The influence of alcohol on sexual assault is a corollary of this unpleasant reaction and there is an evolutionary mechanism at work here too. Behavior that promotes sexual activity, whether or not both participants are willing, tends to produce more offspring. If this behavior has some genetic foundation, it will be propagated. If a taste for alcohol leads to more babies, the instructions for drinking flow down the river of time. Excessive drinking and alcoholism may be extremes of behavior that survive because the underlying genes that make us like alcohol have such a strong tendency to be replicated. The likelihood of becoming a social or an antisocial drinker may come down to the number of copies of specific genes, or lie in variations in these genes. An interesting twist to this idea is that the development of cereal agriculture to provide alcohol allowed ordered communities to flourish in which laws were developed that controlled the excesses of drunken behavior. Even in the most highly developed countries it is clear that this remains a work in progress.

Plants and animals have benefited from yeast's ability to ferment sugars for millions of years. It is an integral player in the ecology of forests, helping animals to find food and plants to advertise their flowers and fruits. There were no problems in Eden until we began dabbling with palm wine, but as we moved from forest to farm we found that alcohol brought equal parts of solace and pain. And while we have many reasons to care about the sugar fungus, it has no deep need for us. Yeast will make a fine living as long as ripe fruits provide it with sugar and animals are drawn to the banquet by the fumes from its splendid fermentations.

3

The Dough Also Rises

Food

In an open patch of forest, beneath the tropical sun, Neolithic hunters gather to slake their thirst with a shared calabash of palm wine, slow-filtered through dry grass to remove insects. Their refreshment is accompanied with wood-fired fish and local fowl, medallions of roasted bush meat—served *au jus*, and a medley of seasonal fruits and wild grains. Omnivorous and well balanced, this diet, along with vigorous daily exercise, kept our ancestors going well into their twenties, but none of them would experience the delights of a pint of bitter and a plate of pub sandwiches.

Yeast fermented wine and beer long before it made bread dough rise. Bread-making without yeast goes back to our beginnings, and grindstones and pestles used to prepare flour from wild plants are even older than the pottery shards and tablets left by Neolithic brewers. These 30,000-year-old implements, found in Europe and Russia, are mottled with starch grains from grasses and ferns.[1] Ancient biscuits baked from these plants must have been very dense, difficult to chew, and wearing on Neolithic teeth. Without salt or any other flavorings, they would have provided little joy beyond the gratification of a full belly. But by the time most humans abandoned hunting and gathering, bread was being baked on a grand scale.

The Egyptians were the first to make the transition from unleavened cereal dough to lighter loaves puffed with the CO_2 from yeast.[2] As we have seen, the fortuitous contamination of dough seems likely to have been the cause, given the abundance of yeast cultivated by brewers in the same communities as the bakers. The swollen dough that resulted would certainly have aroused curiosity, and a little experimentation with the foam, or "barm," skimmed from beer vats would have driven a new wave in bread-making along the Nile. The transfer of a small quantity of starter dough from yesterday's loaves may have been adopted as well.

Bread was baked in homes in the Roman Republic and most families enjoyed unleavened loaves with their meals.[3] Dough was made from a variety of grains, including millet, rye, and barley, but wheat flour made the best bread and supported a huge market for grain imported from North Africa. Romans ate a lot of wheat mixed with milk and boiled to form porridge (*puls*), but unleavened bread was a staple. The practice of home baking began to decline with the formation of the first guild of bakers, called the Collegium Pistorum, in the second century BC. The importance of baking is reflected in the fact that members of this body were freedmen at a time when other tradesmen remained slaves. With the formation of the Collegium, the public ceded control of a food staple and the government began to regulate bread prices. This was a source of considerable discontent during times of shortage, reinforcing the aphorism of the Roman poet, Juvenal, about *panem et circensis*: citizens care little about politics as long as they have bread and the circus.[4]

In the first century AD, Pliny wrote that froth skimmed from beer was being used in bread-making in the western reaches of the Roman Empire.[5] Active principle, or leaven, was also prepared from wheat bran steeped in wine must, which is comparable to the traditional French use of *levain* or a liquid sourdough starter. Pliny made a clear distinction between the dense unleavened loaves baked in Rome and the lighter wheat bread enjoyed by citizens in Gaul and Spain. Leavened breads eventually became popular in Rome as time passed, but bakers using these innovative methods had to compete with the longstanding regard for the healthful nature of heavier loaves. Greek physicians were passionate about the laxative properties of wholegrain bread.

Romans enjoyed many kinds of bread—leavened and unleavened—baked in ovens or over hot coals. *Panis militaris* was a dry biscuit provisioned to the army, *panis boletus* rose in a mold in the

Fig. 8. Loaf of bread fossilized at Pompeii. Displayed at the National Archeological Museum of Naples.

shape of a mushroom, and *panis quadratus* referred to a circular loaf whose dough was divided so that the bread broke easily into chunks. Some Roman bread was preserved in a large oven in a bakery excavated at Pompeii that was filled with charred loaves from the daily bake (Figure 8).[6] These had been scored into eight triangles, making them *panis octagonos* rather than *quadratus*. Several loaves in the Pompeii bakery, as well as bread found in a villa in Herculaneum, had been stamped, "Of Celer, slave of Q. Granius Verus," to indicate its owner. Dough was marked in this fashion at home before it was taken to a communal bakery. The loaves were carbonized by the clouds of superheated air and ash that ripped through the cities during the eruption of Mount Vesuvius. Pliny was dispatched during the same cataclysm, collapsing on the beach down the coast from Herculaneum. Celer fared better, escaping before the pyroclastic surge turned his bread into charcoal, because his name appears in a later list of freed slaves.

Bread was a staple in Medieval Europe, with a daily allowance of two pounds recorded for troops stationed in Scotland in 1300.[7] This provision would have made up most of the daily calorie intake for an infantryman. Nobles preferred lighter bread, which was made from wheat after removing the bran. This "wastel bread" was leavened with ale barm or a starter, using methods that may have been safeguarded in the monasteries after the fall of Rome. Poorer people ate dark "horse breads" made from oat and rye flour, supplemented with rice, peas, and lentils, which is ironic, of course, given the whopping prices paid for artisanal loaves like these today.

The bread sold in the Saturday farmers' market in my town ranges from dense brown bricks to light French baguettes with pure white crumb. The most popular bread occupies the middle of this range—big, chewy loaves whose only serious competition comes from the gluten-free bakery which attracts a growing number of shoppers. Contemplating the current craze for gluten-free bread, I feel like an old Roman, contented with the porridge and flatbread that he has eaten since he played marbles in the street and shaking his head at the young women carrying airy loaves from the bakery. More later about gluten.

The activities of yeast in bread dough are accomplished more swiftly than the fermentation process that underlies the transubstantiation of grape juice into wine. Yeast cells cannot get at the sugars in cereal flour after milling because they are locked in granules of starch. Starch granules are stuffed into its seeds by the cereal plant as an investment in the next generation. When the granules are spread on a microscope slide and illuminated with polarized light, they light up with a Maltese cross pattern. This signature is produced by the arrangement of stacked sheets of glucose molecules. Some of the granules get cracked during milling, but the layered structure is not lost until the flour is mixed with water. Hydration causes the granules to swell, allowing enzymes in the flour to begin chopping the polymer strings into shorter molecules and cleaving sugars from their ends. Wherever this alchemy happens, *Saccharomyces* finds the resulting sludge as inviting as a jug of sweet sap tapped from a palm tree.

The enzymes in cereal flour are proteins called amylases. When these catalysts are hydrated they begin nibbling at the swelling starch granules, releasing sugars. Maltose, which is a disaccharide, or two-sugar combination of glucose molecules, is released by this process, along with solo glucose molecules. Glucose is the ideal food for yeast. It is absorbed through the cell membrane of the fungus and burned as a rich fuel for growth and budding. Fructose is another simple sugar in bread dough that is immediately metabolized by yeast. Yeast cells absorb maltose too, but the fungus gobbles the simpler sugars first. Sucrose, or table sugar, is a second disaccharide that combines glucose and fructose. Yeast snaps this molecule apart at its cell surface in the first minutes of kneading.

When yeast is mixed with cereal dough it begins feeding on the mix of sugars, using the most efficient form of metabolism that consumes oxygen. This aerobic respiration maximizes energy release and the only waste products are CO_2 and water. But stuck inside an increasingly elastic ball of dough, the budding cells become asphyxiated and are forced to shift to fermentation. Fermentation gets less energy from the sugars and reorganizes the carbon skeletons of the sugars into alcohol and CO_2. Unlike its command performance in wine- and beer-making, the formation of alcohol is almost inconsequential in flavoring bread. Yeast makes it in order to extract some energy from sugars in the absence of oxygen and keep growing. Alcohol is merely part of the waste disposal system for the fungus that prevents it from stalling.

In an open beer vat, the CO_2 from fermentation bubbles to the surface where it whips the yeast cells into a white froth. In bread dough, the bubbles inflate and create the air pockets in leavened breads like ciabatta. The gas gets trapped in a network of gluten that develops during kneading. Gluten—the subject of so much dietary interest—is a blend of two proteins called glutenins and gliadins, which work together to produce elastic strings and sheets as the dough is squeezed and stretched, rolled and patted by the hands of the baker. The chemistry is complicated, involving the continuous rearrangement of bonds between the proteins.[8] This allows the dough to rise as the yeast releases gas. Small quantities of lipid molecules in the dough affect gas retention and every part of this intricate manufacturing process can be modified by augmenting the flour with chemicals described as "improvers." Malted barley is a common improver that adds its own amylases to the bread recipe. The extra enzyme speeds up the release of sugars from the starch granules, helping yeast work faster. Some commercial bakeries use a purified form of amylase enzyme produced by another industrial fungus, called *Aspergillus oryzae*.

Baking as an industrial process also requires reliable sources of yeast. Local breweries can meet the needs of small bakeries, but these arrangements are time-consuming and make it difficult to guarantee the uniformity of the loaves. Sourdough starters are more amenable to commercialization, but the complexity of maintaining the same mix of microbes from batch to batch of fermenting dough can make the loaves very costly. Big bakeries need a dependable source of purified yeast that performs the same way, day in and day out. This became evident in the eighteenth century, when the swelling populations of European cities presented a lucrative market for entrepreneurs with the wherewithal to manufacture and distribute live yeast. Dutch brewers and distillers were first on the scene in the 1780s, when they began selling wet yeast cakes, whose sogginess kept the fungus alive.

Production methods were refined in the next century through the introduction of a lever press for squeezing liquid from the cakes and concentrating the yeast cells. The yeast in these compacted blocks came from foam in beer vats. This produced an active form of fresh yeast, but the yields were insufficient to meet the demand. The situation was worsened by the growing popularity of lager beers that were fermented by strains of bottom yeast that were unsuitable for baking. Recognizing this crisis—*panem et circensis*, or else there will be problems—the Federation of Industry of Lower Austria launched a competition in 1847 to increase the yield of purified yeast.[9] The goal was to produce twenty kilograms of yeast from 200 kilograms of grain. The modest prize, split between Dutch guilders and a gold medal, was equivalent to $20,000 in today's currency. A resulting patent would, of course, be worth far more.

The competition was won by Julius Reininghaus, a German chemist, and his business partner, Adolf Mautner, who owned a distillery in Vienna. Their Vienna process began with the cultivation of yeast on a cooled mash of corn, barley, and rye. Removing beer from the equation avoided the bitterness carried over from the hops. Bubbles of gas brought the cells to the surface of the mash, where the foam was collected by skimming, washed with distilled water, allowed to settle, and compressed into cakes. The collaboration between Reininghaus and Mautner marked the conjunction between the industrialization of Europe and the new science of microbiology championed by Louis Pasteur. With our contemporary familiarity with microbes it is easy to devalue Pasteur's experiments showing that bacteria and fungi soured milk and brewed beer, but they helped to sweep away 2,000 years of ignorance.[10] The impact of this intellectual revolution on the practical business of brewing and baking was profound. For the first time, the agents of fermentation were stripped of their mystery and could be controlled.

The Vienna process attracted considerable publicity at the Paris Exposition in 1867, with bread made with Vienna yeast judged "incomparably superior to all other breads." Mautner's factories were producing 6,000 tons of yeast a year in the 1870s, when Eben Horsford visited Europe and prepared a report on Vienna bread for the US government.[11] Horsford was a Harvard chemist who had made a fortune from his own foray into baking, but not by investing in *Saccharomyces*. Indeed, Horsford's interests lay in chemical rather than biological baking. He was the genius behind modern baking powder that is used to lighten cakes, biscuits, and piecrusts with bubbles of CO_2. Before Horsford reformulated the recipe, cream of tartar, which was a by-product of winemaking, had to be mixed with sodium bicarbonate directly before use and the proportions were critical. German chemists had experimented with mixtures of sodium bicarbonate and hydrochloric acid, but the explosive nature of this concoction did not lend itself to making cakes at home. By substituting calcium biphosphate for the cream of tartar, Horsford created a dry premixed powder that any housekeeper could handle.

Baking powder is not suitable for making bread and, lacking the trustworthy source of yeast available to Europeans, American bread-baking remained a hit-and-miss venture. Even bakers in the big cities with access to excellent flour turned out deplorable loaves of uneven texture. Seeking to salvage this industry, Horsford's report on the Vienna process was a wake-up call: the Europeans actually enjoyed their daily bread! This example of foreign supremacy was unacceptable for a nation defined by the belief in its own exceptionalism.[12] The fight was on. And, as has so often proved true in America, the exceptional entrepreneur was an immigrant.

Charles Fleischmann and his brother Max emigrated from Austria to New York in the 1860s, escaping the Austro-Prussian War and arriving at the end of the American Civil War.[13] Charles had worked in the distillery trade in Europe and had intimate knowledge of the latest methods of yeast production. After a short spell in New York, the brothers moved west to Cincinnati, where they began making Kentucky-style bourbon. Cincinnati had the allure of the frontier and hosted a thriving community of Jewish immigrants that welcomed their brethren from Austria. The Fleischmanns operated a liquor dealership from the bustling public landing on the Ohio River frontage for a while, and then moved west of the city to a small town called Riverside where they purchased twelve acres of farmland.

Backed with capital provided by James Gaff, a successful distiller, the Fleischmanns built the first yeast plant in America on the north bank of the Ohio River. They picked the site for its proximity to the railroad line that ran above the plant and access to the great river below. I was surprised to learn that I had driven past the site of their factory on River Road every week for the last twenty years, oblivious to its mycological significance. The brick buildings were demolished long ago and the spot is occupied by a children's baseball park. This unassuming scrap of the riverbank deserves a historical marker as the Birthplace of American Biotechnology.

Adapting the Vienna process to produce cubes of compressed yeast, Charles patented an improved method for collecting froth from fermentation tanks and separating the fungus from the spent cereal malt. After forty-eight hours of growth, the spume was transferred into silk bags, washed in cold water, and the water squeezed from the cells using a hydraulic press. Early marketing targeted German immigrants, but the business expanded beyond this niche very rapidly. Fleischmann and Co.'s Compressed Yeast was advertised to the ten million people visiting the Centennial Exposition in Philadelphia in 1876 through a "Vienna Model Bakery," in which the company demonstrated the production method, baked bread on-site, and seduced the public with Viennese pastries.

Fleischmann's Yeast cakes were protected by patent and the brand was reinforced by a printed warning about imitations on the yellow label of every foil-wrapped cube: "None Genuine without OUR Facsimile Signature *Fleischmann & Co.*"

By the 1890s, the Fleischmanns had become shockingly wealthy. Charles operated four yeast plants, owned vacation properties, a racehorse stable, and a yacht, and was elected to the State Senate. His remains lie in a splendid granite mausoleum, styled as a miniature Parthenon, by the side of a lake in Spring Grove Cemetery, Cincinnati. This is a temple financed by yeast.

Despite its historical significance, the origin and details of the genetic identity of the Fleischmann yeast strain are not known.[14] Popular versions of the story refer to the brothers sailing across the Atlantic with a test tube of live yeast that would make their fortune. This seems unlikely. What they imported from Europe was a familiarity with the latest techniques in microbiology and experience with the Vienna process. The availability of yeast was not a problem. Cincinnati was a center of American brewing in the nineteenth century and there was no shortage of different strains for the fermentation tanks in the Riverside plant.

The Over-the-Rhine district in Cincinnati was populated by German immigrants with more than a passing fondness for beer.[15] Early brewers included Christian Moerlein, from Bavaria, whose label has been rebirthed as one of the great Midwestern craft breweries today. Lager, created using bottom-fermenting yeast, was the most popular beer in Cincinnati in the 1870s. This means that Charles Fleischmann must have worked with one of the brewers who continued to make ales to find a strain of top-fermenting yeast suited for bread-making. Once yeast production began in Riverside, the natural immobility of the fungus may have prevented contamination. Through careful maintenance and breeding from the original stock, it is possible that the strains sold today share a good deal of the original genetic background with the fungi that budded in Cincinnati beer barm 150 years ago.

The original cubes of Fleischmann's Yeast had a very limited shelf life without refrigeration. Although water was squeezed from the blocks of yeast by the manufacturing process, the cells remained hydrated, which allowed them to grow and lose their vigor after a few days at room temperature. This limitation was overcome in the 1940s, when new manufacturing methods provided granules of dry yeast whose animation was suspended until they were mixed with water.[16] A version of this dry yeast with finer granules introduced in the 1980s accelerated the rise, reducing the virtue of patience in bread-making. Neither innovation required any alteration in the identity of the fungus, but the company may have introduced novel strains along the way to suit the various formulations.

Winemakers, on the other hand, have a lot more at stake in their selection of yeast strains, because, as we have seen, the flavor and aroma of their vintages owe a lot to the prolonged labors of yeast. If the complexity of some wines bear reasonable comparison with the magnificence of Versailles, the accomplishments of yeast in a French baguette rise no higher than a sand castle. Testimony to yeast's work comes purely from the shape of the loaf and the texture of the crumb, not from the taste of the bread. Wine yeast can be at it for weeks, whereas the billions of cells revived from the powder in a square sachet can complete their tasks in bread dough in an hour. Following this brief commotion, the heat of the oven exorcizes any fruitiness that lingers from the fermentation of the sugars in the flour.

Commercial yeast was produced in Europe before Charles Fleischmann won over American consumers. Lesaffre, a French company based in Marcq-en-Barœul, opened its first yeast plant in the 1870s and remains the largest yeast producer today. Lesaffre purchased the American brand Red Star Yeast in 2001. Associated British Foods, which owns Fleischmann's Yeast, is second in market share, followed by Lallemand in Canada and Angel Yeast in China.[17] The yeast industry is valued at almost $3 billion and annual production of *Saccharomyces* exceeds two million tons. The increasing popularity of baked goods in Asia is a major factor in the growth of the yeast market and global yeast production is projected to be worth more than $4 billion in 2020.

Modern manufacture of wet and dry yeast for bakeries is a highly technological endeavor in which tons of the fungus are grown and harvested from gleaming stainless steel fermenters or bioreactors. The details of the process are tweaked from plant to plant, which may explain why operators consider the security of their facilities as paramount. While researching this book, most of my calls to big players in yeast production were met with frostiness, and e-mails were ignored. When I did speak with the president of a multinational food corporation, he suggested that there was too much at stake in terms of trade secrets to allow me to shuffle around his floors in sterile booties. I considered donning a white lab coat and goggles and trying to smile my way past the uniformed guards at one of the largest plants in Tennessee, but thought better of the idea when I recalled my failure to smuggle a jar of Marmite into the United States some years ago (an escapade whose significance will be explained later).

Although the details of yeast production favored by different companies remain secret, the basic process is well documented in biotechnology journals.[18] Charles Fleischmann would be awestruck by the scale of modern yeast plants. The production process involves a series of interconnected fermenters of increasing capacity. Yeast is grown on molasses, the syrup that is a by-product from the process of refining sugar from sugar cane and sugar beet. At the beginning of the process, the yeast is supplied with a single shot of molasses and grows and divides until it begins to deplete the sugars and run out of oxygen. To counteract the stagnation of the yeast, the suspended cells are flushed with oxygen and supplied with more food. This fed-batch method of fermentation provides the yeast with the optimal conditions to maintain growth and budding. As they outgrow one fermenter they are streamed to the next in line. If all goes well, successive harvests of 10–20 tons of wet yeast can be filtered from commercial bioreactors with a capacity of 150,000 liters. Centrifuges and presses are used to concentrate the fungus before it is formulated as cream yeast or dried granules.

Molasses are the best food for yeast, but they do have some drawbacks. One is the rising cost of this raw material, driven, in part, by its use to grow yeast in bioethanol plants. Another problem is that molasses from some sources can be contaminated with herbicides, insecticides, fungicides, and toxic heavy metals. All of these chemicals can upset yeast as it attempts to grow. The solution is to combine molasses from different sources so that the level of any particular toxin is diluted in the resulting blend. Before molasses are fed to yeast they are supplemented with vitamins and with urea that serves as a source of nitrogen for the fungus. Even with all of this careful husbandry the fungus shows signs of metabolic stress throughout the fermentation.[19]

There is little that is carbon-neutral about yeast production in an industrial setting. It is an energy-intensive manufacturing process that relies on petroleum-based agriculture and produces huge quantities of contaminated wastewater. The selection of the Riverside site for America's first yeast plant outside Cincinnati was encouraged by the availability of fresh water for manufacturing and waste disposal. Effluent from the Fleischmann plant was piped directly into the Ohio River, along with the waste from nearby slaughterhouses, tanneries, and the chemical industry.

Wastewater treatment is a major cost for today's yeast industry. Some corporations are working with water purification experts to develop cleanup systems that rely on other microorganisms to ferment the organic waste left behind by yeast. Environmental laws are driving this new technology, but the continuing reliance of the yeast industry on river access is telling. An aerial survey courtesy of Google Earth shows that the Memphis plant operated by Lallemand is spitting distance from the Mississippi River; AB Mauri's huge yeast plant in Veracruz is just above the Rio Blanco that drains into the Gulf of Mexico; and the world's biggest yeast plant, owned by Angel Yeast, is in Yichang, a port city on the Yangtze. The Yichang plant produces 500 tons of yeast a month.

These massive industrial plants exemplify the extent to which food manufacturing has pushed all of the non-human contestants beyond their natural physical limits. Nobody is going to feel sorry about the working conditions of *Saccharomyces*, but yeast has been tormented in a fashion reminiscent of battery chickens. Rather than breeding yeast that makes more buds and makes them more swiftly, the strategy in the food industry has been to crowd the fungus in an environment in which it is compelled to work quickly. This principle applies to the process of manufacturing masses of the fungus to be shipped to bakeries, as well as its subsequent labors in bread-making.

Efforts to improve upon traditional baking practices culminated in a major research initiative in the 1950s by the British Baking Industries Research Association based in Chorleywood in Hertfordshire. The aim was bold: to invent the most cost-effective way to produce the perfect loaf of bread. The Chorleywood bread process resulted from this endeavor, accelerating dough preparation and allowing bakers to make use of lower-quality wheat flour.[20] Food scientists modified existing recipes for white bread and experimented with mechanical mixers to minimize the time between combining the ingredients and packaging the loaves. The solution included tripling the amount of yeast at the start, adding some vegetable fat, and mixing the dough with unprecedented ferocity for three minutes. This is a good deal faster than kneading dough on a floured kitchen counter.

In the Chorleywood process, flour from hybrid wheat is milled to a fine powder to facilitate the absorption of water. After mixing, the dough is heated and subjected to changes in pressure to force air into the dough and then control bubble-formation from the overexcited yeast cells. The fungus is given 45–50 minutes to raise the dough and baking is done in as little as seventeen minutes. Even with a subsequent two-hour cooling period, the loaf is sliced and in the bag in 3½ hours. Before the introduction of the Chorleywood process, British bakers allowed yeast to multiply and raise the dough in an overnight proofing. The addition of more yeast at the beginning of the new process made the wait unnecessary.

Scientists at Chorleywood were afforded a free rein and generous budget to support their pursuit of the perfect loaf. Descriptions of the project provide a sense of the necessity of intensive engagement of individual scientists, as well as effective teamwork.

The investigators mixed and baked thousands of batches of dough and photographed rows of bread slices to record their size and shape. This seems like astonishingly tedious work to outsiders, but it is surprising how exciting science can become when a big group is working toward a common goal. Making the perfect bread is not the same as searching for new antibiotics, but the experience of the ins and outs of lab work is often very similar for scientists working on completely different things.

One of the instruments used by the Chorleywood scientists was the Chopin Alveograph. The Alveograph was invented in 1920 and has remained an indispensable tool for commercial bakers.[21] The modern version is a computer-controlled marvel that provides data on the extensibility of the dough by inflating biscuit-sized samples into perfect white balloons inside a temperature- and humidity-controlled chamber. A key measurement is made when the expanding balloon pops. Balloon size is related to the extensibility of the dough. Watching the balloons grow and collapse through the glass window into the chamber is hypnotic. The instrument uses pressurized air for inflation; the natural process is driven by billions of yeast cells exhaling puffs of CO_2. With a price tag of $70,000, the automated Alveograph is out of reach of most home bakers. This technology has evolved alongside the development of industrial baking and supports bakeries that produce hundreds of thousands of loaves per day.

The Chorleywood experiments come from an era of splendid investment in research by the food industry. In this case, the corporate goal was straightforward and resulted in an immense payback for British baking. Some food companies supported first-rate research on microbiological problems far removed from their products. They had faith, I suppose, in the possibility of long-term pay-offs, but it is difficult to see how some projects were funded at all. A classic paper from 1958 on the capture of microscopic nematode worms by soil fungi—revered by the three or four biologists who study this mechanism today—came from Symbol Biscuits Limited in the Lancashire town of Blackpool.[22] The juxtaposition of a young researcher triggering microscopic nooses set by a fungus, in a lab next to workers in hairnets picking cookies from a conveyor belt, belongs to the Theater of the Absurd. But this did happen, and the scientist went on to become a professor of food science at the University of Leeds. With increasing globalization and a narrowing of commercial investment in long-term research, options for corporate PhDs withered in the 1980s and science is less inventive for it.

Chorleywood has become the industry standard in the United Kingdom, and has been adopted in many other countries. The wheat grown in North America tends to have a higher gluten content than European wheat, and the resulting dough is resistant to high-speed mixing. For this reason, American bread is made by batch mixing, in which the fungus is given the luxury of 2–4 hours of proofing, which means allowing the dough to sit and ferment before baking. Wonder Bread has been a popular mass-produced loaf in North America since the 1920s. It is marketed as "soft, white, and as wholesome as childhood itself," which is unsettling from numerous perspectives.[23] The bakers in my farmers' market view this product with the horror that oenophiles show toward fortified wines sold in American liquor stores with names like "Night Train Express" or "MD 20/20," which sounds like an insecticide. Sliced white bread and wine with artificial color are sorrowful distortions of the historic collaboration between mankind and *Saccharomyces*.

Conspiracy theorists have claimed that bakeries have developed Frankenyeasts that perform foul deeds in modern bread-making.[24] In the Chorleywood version of this story, the introduction of a modified yeast strain is linked to a purported increase in celiac disease. This seems very unlikely given the gluten sensitivity in people who suffer from this illness, because Chorleywood was designed to produce low-gluten wheat bread. Manipulation of yeast strains is certainly the concern of yeast biotechnologists, but it is more likely that any health problems associated with modern bread come from enzymes and other chemical additives.

Sourdough bread is marketed very differently from the white sliced loaves sold in plastic wrappers and it is created by a complex mélange of microbes rather than a monoculture of *Saccharomyces*. The starter or "mother dough" for these loaves is a pasty white mix of fermenting cereal flour that combines different species of yeast with lactic acid bacteria.[25] When sugars are plentiful and oxygen levels dip, the bacteria in sourdough produce lactic acid and CO_2, rather than the ethanol and CO_2 released by yeast. Similar biochemistry applies to the buildup of lactic acid in our muscles when we deplete oxygen during strenuous exercise. The lactic acid causes the burning sensation associated with muscle exhaustion. Acetic acid is a second by-product of some of the lactic acid bacteria that contributes to the taste of sourdough.

Lactic acid bacteria work alongside yeasts to create sourdoughs. The bacteria are species of *Lactobacillus*, including *Lactobacillus sanfranciscensis*, which was discovered in the 1970s by microbiologists studying the origin of the unique flavor of bread in the Bay Area in California. *Lactobacillus sanfranciscensis* was new to science and it was associated with a yeast identified as *Candida milleri*. The bacterium acidifies its surroundings for two reasons. First, because the process allows the cells to continue sugar metabolism, and second, because it is an effective way to get rid of competitors. These are, of course, precisely the same benefits obtained by yeast when it generates alcohol. *Candida milleri* is an acid-tolerant fungus, suiting it for life with the sourdough bacterium, and the relationship is solidified by the inability of the yeast to consume maltose, which is favored by the *Lactobacillus*.

Despite its name, *Lactobacillus sanfranciscensis* is not endemic to California, but serves as a natural participant in sourdough baking all over the world with *Candida milleri*. With the knowledge of this microbial marriage, it would seem easy as pie to make San Francisco-style bread anywhere. This does not work, however, because other kinds of lactic acid bacteria share the space with this pair of microbes. The complex mixtures of bacteria and yeasts used in some Bay Area bakeries have been nurtured for many human generations of bakers. The famous Boudin Bakery claims that its starters have been maintained since the 1840s, when Isadore Boudin borrowed a mix from miners participating in the Gold Rush.

Industrialized approaches to making sourdough bread have come to rely on higher temperatures for fermentation of the starter, and these conditions are too hostile for *Lactobacillus sanfranciscensis* and *Candida milleri*. These breads, sold in national grocery chains, are made by other bacterial species and *Saccharomyces cerevisiae* is added to compensate for the loss of the original fungal participants.

The microbiology of sourdough starters used to make sour-tasting bread is not the same as a levain used to make bread according to the French tradition. With a levain, there is no need to add purified yeast to the new batch of dough, but this method does not necessarily produce bread with a sour taste. This can be confusing, because sourdough starter and levain are often used as synonyms. If the stereotypical baker who makes sliced white loaves in a factory resembles a technician in a biowarfare lab, her counterpart in the sourdough business is likely to be a bearded dude with muscular forearms who grabs his surfboard when the morning bake is done. Many do not care for sour flavors, but the current fashion for artisanal sourdough loaves can overcome the mere obstacle of taste.

The significance of yeast in the food industry extends to many products beyond the bakery. It plays a starring role in the $100 billion global market for chocolate. This rainforest commodity is made from raw cocoa pellets, called nibs, ground from roasted seeds. Cacao is the preferred spelling that comes from a Spanish noun, but cocoa is less jarring to English speakers and is used widely. The production of nibs requires the work of a succession of yeasts and bacteria that reduce the astringency of the seeds when they are extracted from their pods.[26] Astringency is sometimes confused with sourness. An astringent taste is caused by tannins that bind proteins in saliva, making the mouth feel dry and rough. Sourness comes from acidity. When cocoa seeds are spilled from their pods, they are surrounded by white pulp. This pulp-bean mass is arranged in heaps, boxes, or baskets and allowed to ferment in the open air for up to a week.

The first yeasts that show up have names that cannot be read out loud without a few test runs—*Hanseniaspora guilliermondii* and *Hanseniaspora opuntiae*—followed by multiple strains of *Saccharomyces* and the utterly unpronounceable *Pichia kudriavzevii*, also called *Issatchenkia orientalis*.[27] The tongue twistiness of the latter species qualifies this fungus as the Engelbert Humperdinck of mycology. Besides its taste for cocoa pulp, *Issatchenkia* has been isolated from soil, cabbage waste, and, unsettlingly, from human pus, sputum, and feces.[28]

Saccharomyces is present throughout pulp fermentation but held in check by the acidic conditions created by bacteria. As different populations of microbes blossom, they feed on the citric acid formed in the early stages of the fermentation and create an opening for *Saccharomyces. Saccharomyces* responds by budding and, letting no good deed go unpunished, rewards its predecessors by poisoning them with alcohol. This rise and fall of microbial populations is very similar to the ecological succession in palm wine. DNA fingerprinting has been used to study the waves of microorganisms that wax and wane as the conditions change in the bean boxes. The first author of a study of Malaysian beans was Zoi Papalexandratou, who works for a company that supplies cocoa from Nicaragua.[29] I hope that one of her protégés does not discover a new yeast and name it in her honor.

Once the fermentation of the pulp is complete, the seeds are dried, their thin husks are removed, and the remaining tissue is ground to produce the nibs. These are combined with other ingredients to make chocolate. Research on cocoa fermentation is being taken very seriously by chocolate producers who would like to take some of the guesswork out of the process. If investigators can identify strains of yeast and bacteria that are the most effective at removing astringency from the nibs without sweeping away their glorious flavors they may make a fortune. Unlike bad wine, however, even the cheapest chocolate bar is pretty good, and it is going to be very difficult to make the best stuff taste even better. Attention is also turning to yeast strains associated with the fermentation of coffee cherries and the underlying pulp. *Saccharomyces* is always part of the mix and the coffee strains are very different from the cocoa yeasts.[30] Their role in flavoring the beans before they are roasted is not known, but is causing a lot of excitement among coffee producers and consumers.

The popularity of the Hershey Milk Chocolate Bar, introduced by the Hershey Chocolate Company in America in 1900, stimulated the Fleischmann Company to diversify its role in the food industry. The concept was to market cakes of compressed yeast as a delicious and healthy alternative to candy bars. And why not? These blocks of yeast were smooth and rich, and, unlike chocolate, had the added benefit of curing all illnesses.[31] Here is one verse from an advertising ditty from the 1920s that will suffice as an illustration of the Fleischmann promotional strategy:

> Children love this lovely
> Creamy food delight
> Let them eat it daily
> Every morning, noon and night
> You will see them growing
> Stronger every day
> Taking yeast this handy
> Dandy candy way

This was part of a "Yeast for Health" ad campaign that went on to set new lows in ridiculous product claims. One advertisement published in *Popular Mechanics* magazine showed a teenage boy with acne—the photograph makes him look like a smallpox victim—whose prospects with the fairer sex improve greatly after he consumes three blocks of yeast a day to remove "waste poisons in the blood."[32] Another advertisement recommended springtime as an auspicious season for an intestinal cleanse with Fleischmann's Yeast, "a natural remedy for constipation and the ills that go with it." This therapy was promoted by "Dr. Ploos van Amstel, famous intestinal specialist of Amsterdam, Holland," who is pictured as a bald clergyman who purports to be unacquainted with anything "as satisfactory as YEAST," to "cleanse the system." A second physician, Dr. Henri Vignes of Paris, was similarly evangelical about the power of yeast to combat "sluggish intestines," and implored his readers to consume "three cakes every day!" In the photograph in the advertisement, Dr. Vignes seems to threaten anyone who ignores his prescription (Figure 9).

"When intestines are *sluggish* I prescribe Fresh Yeast..."

reports the noted
DR. HENRI VIGNES
of Paris

Fleischmann's Yeast is fresh yeast... the only kind that benefits you fully. Eat three cakes every day!

Fig. 9. Details from Fleischmann's yeast advertisement published in 1931.

The Federal Trade Commission was concerned by these untested assertions, but profits for the Fleischmann Company soared. They found a new way to market their products by sponsoring "The Rudy Vallée Show" on NBC Radio in the 1930s. With sponsorship came naming rights, and the show was also known as "The Fleischmann Yeast Hour." This musical variety show introduced Milton Berle, and comedy duo George Burns and Grace Allen, and became the first radio program to be hosted by an African American when Louis Armstrong was guest presenter in 1937.

In a more serious vein, plain yeast has been considered as a solution to human hunger since our species passed the two billion mark, or thereabouts, early in the twentieth century. In those days, when there were five billion fewer of us than now, economists showed greater interest in the prospect of a Malthusian nightmare of mass starvation. We know now, of course, that there are no limits to the wonders of population growth, the economy is better for it, genetic modification of crops is a miracle for the poor, the environment is cleaner than ever before, and the climate is holding steady. In the last century, however, famine dispatched five million Russians in 1921, seven million Bengalis in 1943, and forty-three million people in China between 1958 and 1961. Lesser famines ravaged Ukraine, Vietnam, and North Korea. Even if politics was the primary cause of these catastrophes, food shortages have a way of sharpening interest in human demographics and agricultural capacity. Raw yeast has seemed to be the answer to these challenges on a number of occasions.

As a food, yeast is an example of a single-cell protein, distinguished from meat and fish that are multicellular proteins. The term "single-cell protein" was coined in 1966 by Carroll Wilson, a professor at MIT, who is better known for his analysis of the environmental consequences of relentless economic growth.[33] It was far superior to the baffling competitor, "protéins de biosynthèse," suggested by the French Académie nationale de médecine, which does nothing but restate the fact that proteins are biological products. The use of single-cell protein as an animal feed was explored in Germany during the First and Second World Wars and scientists harnessed the power of new methods of fed-batch fermentation to optimize yeast production. Brown sugary waste from the manufacture of paper pulp, called sulfite liquor, was used to feed the yeast, which was separated as a cream from the fermentation tanks. *Saccharomyces* could be grown in this fashion, but another yeast, *Candida utilis* or "torula yeast," was better suited for this work.

The story of food yeast took an horrific turn in the 1940s, when the Nazi *Schutzstaffel* (SS) became interested in using it to make protein-rich foods to fortify frontline troops.[34] In 1942, Hans Kammler, an SS officer trained as a civil engineer, supported the construction of a yeast plant using slave labor in Wittenberge in northern Germany. This was a low-priority project for the Nazis, but Kammler struck a deal with a private manufacturing company that agreed to funnel 75 percent of the yeast produced in the plant to the SS. Kammler soon lost interest in this venture, but Heinrich Himmler recognized the importance of new methods of food production as the war continued.

Rather than risking the health of the troops, the SS conducted experiments on the nutritional value of yeast in their concentration camps. Eduard Wirths, chief SS doctor at Auschwitz, fed starving prisoners a mixture of fodder yeast and nettles, and Ernst-Günther Schenck, inspector of nutrition for the SS, investigated the effects of a sausage made from yeast at Mauthausen-Gusen. Hundreds of prisoners starved to death, the Nazis photographed their wasting bodies for their archives, and many survivors of the experiment were gassed. As the supervisor of Joseph Mengele, Wirths presided over experiments still darker and hanged himself in 1945. Schenck was captured by the Russian army in 1945, and released in 1953. Through the remarkable massaging of evidence that kept many former Nazis free, he evaded imprisonment and died at age 94 in Aachen.

One of the ironies of the Nazis' convictions about racial superiority is the fact that they banished many of the brightest scientists of their generation. A number of them were experts on yeast science.[35] Gerty Cori, whose Jewish heritage prevented her from working at the University in Prague, moved to Buffalo, New York, with her husband Carl in 1931. Gerty and Carl were awarded a Nobel Prize in 1947 for their research on sugar metabolism. Otto Meyerhof won a Nobel Prize for his work on lactic acid metabolism in 1922 and left Berlin for the United States in 1940. Carl Neuberg, who was an expert on alcohol fermentation by yeast, was forced from his job in 1934 and fled Germany in 1939. Neuberg was not awarded a Nobel Prize, but he is known as the father of modern biochemistry, which is not too shabby. His contemporary, Otto Warburg, who did win a Nobel Prize, was the exception in this list of scientists because he continued his research in Berlin throughout the war. Warburg, whose father was Jewish, was reevaluated as one-quarter Jewish and protected by Hermann Göring, who lived by the maxim attributed to a Vienna mayor, "I will decide who is a Jew." The Nazis did not want to lose Warburg because his experiments were pertinent to cancer research and Hitler had become very fearful of this emperor of maladies. Warburg's research assistant, Hans Krebs, who was not favored in the same fashion by the Nazis, moved to England and won the 1953 Nobel Prize in Physiology or Medicine. The list of exiled talent is endless.

Pursuing the same interest in growing yeast as a single-cell protein for human consumption, as well as a cheap animal feed, British Petroleum developed the technology to grow yeast on paraffin generated in oil refineries in the 1960s.[36] Oil prices were exceedingly low at this time, less than half the cost per barrel today, even after correcting for inflation. Yeast fermented this energy-rich food in a new kind of fermenter called the "air-lift" that injected pressurized air into the stainless steel tanks to circulate the cells. With massive investment by petrochemical companies, factories dedicated to raising single-cell protein opened in Europe, the United States, and Japan. The Soviet Union also adopted the technology with great enthusiasm and established the All Union Institute for Synthesis of Protein to administer the growing number of yeast plants.[37] The potential danger of consuming traces of petrochemicals harvested with the yeast prevented the use of this food source for human consumption, but the yeast was incorporated into animal feed and provided the calories for farm-raised fish.

The US became very interested in increases in food yeast output in the USSR and commissioned a top secret report by the CIA in 1977.[38] This document was released to the public in 1999 with redacted paragraphs that might, if unredacted, explain why the CIA was so concerned. The strategic advantage for the USSR of food yeast as a buffer against a domestic or international agricultural crisis provides a likely explanation. The CIA need not have worried. Investment in the process fizzled out as concerns grew regarding the safety of incorporating petroleum fractions into the food chain and as oil prices rose during the 1970s energy crisis.

Plants producing petrochemical yeast closed or converted to other fermentation methods.

While today's market for food yeast for human consumption is limited to its promotion as a vitamin-rich dietary supplement sold in health food stores, fodder or feed yeast grown on molasses generates more than $300 million in global sales.[39] A variety of yeast species are incorporated into animal feed. *Candida utilis* is the most important of these and is used in cattle and poultry feed. *Saccharomyces* is fed to dairy cows for its proven effectiveness in reducing heat stress in the animals and boosting milk production. The mechanism underlying these benefits is unclear, but *Saccharomyces* may work in the digestive system of the animals by consuming sugars in the rich diet of hay, silage, and blended animal feeds. Cows gulp air during feeding, which is a liability in the rumen where cellulose digestion depends on anaerobic conditions. Yeast added to the feed may offset this effect by removing oxygen as it begins to metabolize the sugars.

Saccharomyces is not a normal part of the rumen microbiome and is considered a probiotic by dairy farmers. The natural residents include species of Neocallimastigomycota, whose name echoes an immensely annoying song from *Mary Poppins*. These strange microbes are poisoned by oxygen and use a hydra-like wig of beating cilia to swim around the fluid in the rumen.[40] When *Saccharomyces* gets to work on the sugars in the rumen, it releases alcohol as the oxygen levels fall. Does the trickle of alcohol enhance the mood of the animals at all? That might be too much to wish for on behalf of dairy cows.

A final example of the food uses of *Saccharomyces* brings me back to my Marmite escapade. Like many Brits, I was weaned on Marmite, known as Vegemite to those who live Down Under.

Marmite and Vegemite are thick slimes of yeast obtained from breweries and salted to create a spread that is shunned by most of humanity.[41] It is the British version of natto, the Japanese paste of fermented soybeans. Marmite tastes nothing like natto, but, like the Japanese food, is judged repulsive by most people who were not introduced to it in infancy. Marmite is yeast unadorned, the essence of the fungus, extracted in all its glory and sealed in little jars. Marmite is inexpensive in the United Kingdom, but sold at inflated prices in the United States where it is labeled as a "specialty food." Unilever, which makes Marmite, brings out limited editions of the stuff from time to time. In a triumph of reason, they made a batch of Marmite from Guinness Yeast, "Alcohol Free/Limited Edition," whose rarity justified squirreling a jar back into Cincinnati. Agents with the Transportation Security Administration were unimpressed by my homily on free trade and tossed it into a box filled with miscellaneous contraband.

People of some religious faiths have misgivings about yeast. Jewish anxieties are apparent in the Old Testament, which threatens people who eat leavened bread during Passover with the Hebrew equivalent of excommunication. Leavened food, referred to as *chametz*, must be removed from Jewish homes during Passover and the observant are required to discard every crumb of bread, every morsel of food transformed by yeast. There are some workarounds, including the sale of *chametz* to gentiles via the brokerage of a Rabbi, to be repurchased after the holiday. This may seem frivolous, but the practice seems to be rooted in the association of leaven with corruption, and unleavened bread's representation of sincerity and truth. In a related vein, Roman Catholics and many Protestant churches mandate the use of unleavened bread and wafers in the celebration of the Eucharist and Communion.

The Eastern Orthodox and Eastern Catholic churches swing the opposite way, forbidding unleavened bread and favoring leavened bread as a symbol of the New Covenant with God that will see fulfillment after the Second Coming.

In very small print on the labels of special jars of Marmite are claims that the paste is a Kosher product, but there is some disagreement within rabbinical circles about whether this counts during Passover.[42] This seems unnecessary, based on my insights about *chametz*, because only the essence of yeast is sold in the jars and this is incapable of leavening anything other than one's temperament.

4

Frankenyeast

Cells

If the human relationship with yeast went no further than our alliance in brewing and baking, the sugar fungus would have little competition for the status of civilization's greatest microbial ally. Yet, in our time, its significance has expanded as yeast has taken center stage as the marvel of biological research and biotechnology. Part of the reason for its scientific celebrity lies in its simplicity. Yeast is a streamlined expression of life. This makes the fungus a superb subject for experimentation. We know how it harvests energy from sugars; we have scrutinized the birth, life, and death of its cells at the molecular level; and, now that its genome has been sequenced, we are edging closer to being able to control everything this microbe does. Far from a trivial exercise, the science of yeast offers a rich illustration of the workings of biology—and the meaning of life.

Yeast is built like a human cell, with a central nucleus, and uses a compact genetic blueprint that lends itself to experimental manipulation. Another experimental attribute of yeast is its short generation time. Each cell produces a bud every 1–2 hours, which means that a squirt of yeasts from a pipet tip into a broth will yield billions of individuals with the same genetic makeup—or clones—in an overnight culture. These features make it an ideal model for biological investigation. *Saccharomyces* has been the subject of intensive research for almost 200 years and thousands of scientists are engaged in studying its biology today. Tens of thousands of research papers have been written about this single-celled fungus, and colossal sums of money have been spent on its manipulation in the lab. This effort has been profitable by any measure. We know more about yeast than about any other organism whose structure is more complicated than a bacterium. This knowledge has been translated into advances in medicine and biotechnology, the food industry, and brewing. More importantly, biofuel research on yeast may be the last best hope in our era of climate change.

Modern civilization is undoubtedly richer for this scientific enterprise. And yet, it is impossible to fully comprehend the inside of a yeast cell or to imagine what goes on within this speck of life. Anyone who has read a popular science article about the origin of the universe or discussed the nature of nothingness over a few ales has bumped up against the unimaginable. Students of nature, young and old, recognize that some things remain beyond understanding. After all, there was nothing in our evolutionary history to advantage familiarity with the equivalence of mass and energy, and the continuum of space and time. This provides us with an excuse for feeling confused about relativity, but the unknowability of a little thing like yeast, this relatively simple expression of life, seems more disturbing.

The challenge does not come from a lack of cell biological knowledge. Even Nobel laureates who have dedicated themselves to the study of yeast cells have been daunted by the enormity of this task. The fundamental problem is not limited to our comprehension of yeast. Biology teachers misrepresent the cells of every other kind of organism when they explain the structure of this foundational blob to students. The classic diagram of a eukaryotic cell shows the outer boundary, or cell membrane, enclosing a nucleus, a scattering of burrito-shaped mitochondria, a slice of Golgi apparatus, associated membranes, and ribosomes represented as dots. The nucleus is the defining feature of the eukaryote cell, in which the chromosomes are penned, rather than free as we find in the simpler prokaryote cells of bacteria (Figure 10). This portrait is a long way from reality. It would be easier to use Giacometti's anorexic bronzes to explain the intricacies of human anatomy than to bridge the gulf between the diagram of the cell and the living thing.

To begin to appreciate yeast, we need to think about how it works. We have solved the chemical structure of most of the components of the yeast cell and know how many of them are organized inside the cell. The problem comes when we try to imagine how the different parts come together and do the things that yeast does.

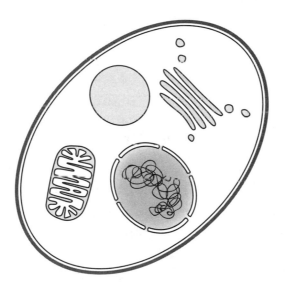

Fig. 10. Simple diagram of a eukaryotic cell.

How, for instance, do the twelve distinct enzymes involved in making alcohol from glucose accomplish this vital task?

Standard illustrations of the metabolic pathway begin with an arrow from glucose to glucose-6-phosphate, and show that this first reaction is catalyzed by an enzyme called hexokinase. In subsequent reactions, the sugar molecule is reorganized, split in two, CO_2 is released, and energy is harvested. The last enzyme in the sequence, alcohol dehydrogenase, yields the yeasts' defensive chemical and the source of our consolation (see Figure 3). Biochemistry students learn these pathways when they study cellular metabolism, and consult diagrams showing tangles of chemical transformations that are beyond memorization. Nine hundred genes encode enzymes that accelerate 1,888 reactions in yeast cells producing more than 1,400 kinds of molecule.[1] Many of these reactions take place all the time. The moment a hexokinase protein releases one modified glucose molecule, it locks on to the next one and attaches a phosphate group to form glucose-6-phosphate. Each yeast cell contains 123,000 copies of one type of hexokinase (it has two versions of the enzyme), so an awful lot of glucose is changing hands. Considering that this is one of numerous concurrent chemical processes, driven by tens of thousands of copies of hundreds of enzymes, we can begin to appreciate the terrific agitation inside the cell. Millions of metabolic reactions keep the cell alive.

The biochemistry of alcoholic fermentation can be likened to farmers putting out a barn fire by handing buckets of water along a chain. The buckets are the chemical intermediates and the farmers are the enzymes. But the analogy falters when we scale up to the chemistry of the whole cell, which requires hundreds of chains of firefighters, some of them passing buckets between chains, others shouting instructions to speed up, slow down, or stop for a moment. A flamenco dancer may be closer to capturing the spirit of these reactions, stamping the floor at an insane speed but never losing control of the precise choreography.

With whatever metaphor we adopt, it is impossible to catch more than a fleeting vision of the living cell beyond the confusion of lines in metabolic diagrams. The apparently effortless manner in which yeast keeps on feeding, releasing CO_2, producing ethanol, and making buds obscures the enormity of the business of being alive. The little yeast is more complicated, physically, than any known extraterrestrial entity. Consider the sun. It is very big and very powerful, but runs on the very simple chemistry of hydrogen atoms fusing to form helium in its plasma core. Compare that with the myriad chemical reactions occurring at every instant in a yeast cell, and the labor of our slowly dying star is eclipsed by the magnificence of life.

The reactions of alcoholic fermentation, protein synthesis, DNA copying, and so on, happen in the gooey interior of the cell (Figure 11). Seventy percent or more of a yeast is water, which serves as the solvent in which other molecules dissolve. Anything that is not tethered to a membrane, attached to a protein filament, or linked with another big assembly of molecules like a ribosome is free to diffuse on its own through the water.[2] Diffusion is very effective in such a small space, allowing every individual molecule to meet all of the other molecules dissolved in the cell every few seconds. This means that a glucose molecule absorbed by the cell bumps into a hexokinase enzyme almost instantaneously.

The subsequent reactions of sugar metabolism occur automatically because the other enzymes in the reaction chain are floating around too. There is some evidence that the process is accelerated by the clustering of enzymes that are involved in the same sets

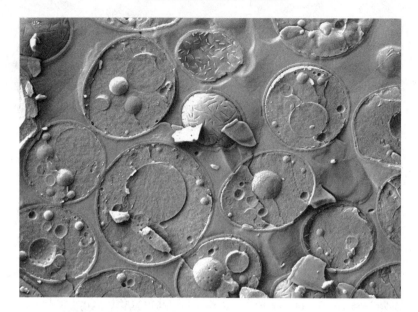

Fig. 11. Electron micrograph of freeze-fractured yeast cells showing spherical vacuoles and single nucleus in the cells fractured through their granular cytoplasm, and the cell wall and cell membrane in the pair of cells in the upper middle of the image that fractured at the cell surface.

of reactions. One way that this may happen is through the corralling of enzymes by strands of protein that form a scaffold inside the cell.[3] This meshwork of filaments and tubes is called the cytoskeleton. The cytoskeleton performs a lot of functions, mostly mechanical, forming guides for the movement of little spheres, or vesicles, that carry materials to and from the cell surface, controlling the division of the nucleus, and remodeling the cell as it forms a bud.[4]

If we can allow ourselves to slip into a Zen-like appreciation of the perfect curves of a yeast cell, the plain fungus is refashioned as a supermodel. The scars from budding are the marks of experience, of a life well lived. The absence of stretch marks around these blemishes is a consequence of the inflation of the yeast cell to around two atmospheres of pressure, the same as a mountain bike tire. This pressure comes from the absorption of water by osmosis. Water moves into yeasts across their membranes because they contain more dissolved molecules and ions than their surroundings. As concentrated blobs of matter, yeasts have no problem keeping hydrated and preternaturally smooth in fluids like grape must and wet dough.

The silky exterior of the yeast is furnished by the cell wall, which is a composite of proteins, carbohydrates, and tough strings of a molecule called chitin (which also forms the external skeletons of insects). At one time it was thought that the cell wall was structured as a series of layers, comparable to plywood or a laminate floor. This proved too simplistic and a reinforced polymer, like fiberglass, was adopted as a better facsimile. Either way, the wall around the yeast cell is an astonishingly complicated thing. Cell biologists have certainly learned a lot about the constituents of cell walls of yeasts and other fungi in the last decade or two, but the goal of a complete account seems farther away than ever. More research produces more data, yet the wall, as is true of the whole cell, escapes our grasp. When a beam of electrons sweeps over the surface of a yeast cell in an electron microscope, the scattered particles are processed to form a three-dimensional image of the cell surface: gray, smooth, curved, and unknowable still.[5]

Beyond the chemical composition of the wall, there are a host of unanswered questions about the way it behaves as a physical material. How is it stretched as the cell grows and buds? How does it accommodate more chitin without being ruptured by the internal pressure that pushes against its inner surface? Beneath the wall is a cell membrane, a mosaic of fats and proteins that serves as the gatekeeper between the environment and cell, absorbing food, and releasing waste materials. Together, the membrane and wall prevent noxious compounds from getting into the cell and both surface layers expand in perfect harmony as the cell grows.

Like the wall, the membrane is supremely complicated. Proteins that act as channels sit in membranes and funnel ions of calcium (Ca^{2+}), sodium (Na^+), potassium (K^+), chlorine (Cl), and other elements that pass to and fro maintaining the balance of salts in the cell. Ions carry electric charges, and when they flow through a channel protein they transmit a tiny current across the membrane. In an astonishing piece of work published in the 1980s, researchers described how they had been able to detect the flux of ions through the yeast membrane.[6]

The technique used to measure electrical currents flowing through yeast membranes is called, somewhat plainly, the "patch clamp." To patch clamp a yeast, the wall is dissolved from the cell surface using enzymes, and an open glass needle, or pipette, is attached to a spot on the naked membrane. The experimenter views this procedure through a microscope, guiding the movement of the pipette with a micromanipulator that scales down hand movements that would otherwise obliterate the cell. When the tip of the pipette contacts the yeast, suction is applied by a delicate "kiss" to the end of a plastic tube that is connected to the interior of the pipette. When the kiss is returned by the cell membrane, a very tight seal forms with the glass. This isolates a minuscule patch of membrane, enabling the motion of ions across it to be recorded as steps in current using an exceedingly sensitive amplifier.

Unappreciated as a living thing until the 1830s, yeast can now be probed down to the level of the movement of single atoms across its membrane. The reason for investing so much time and money in understanding yeast can be appreciated on a number of levels. Because yeast is a good model for understanding how our cells work, the medical ramifications of the experiments on the sugar fungus are never far from view. This is the predominant justification for public investment in yeast research. Private money is attracted by the more immediate uses of yeast in brewing, baking, and other biotechnological enterprises. These applications of the science fund the research, but do not come close to explaining why scientists spend their lives in the lab. The strongest motivation is the thrill of the chase, the drive to grasp the secrets of the cell that have eluded previous investigators.

In 1996, the study of yeast was revolutionized by another publication, a short article in *Science* titled, "Life With 6,000 Genes."[7] It reported the completion of the project to sequence the entire genome of *Saccharomyces*. The bacterium *Haemophilus influenza*, which causes meningitis in children, had been sequenced in the previous year, but yeast was the first eukaryote to be exposed in this fashion. The labor involved in this project, and the technological feat, were breathtaking: *Haemophilus* has 1.8 million As, Ts, Gs, and Cs; yeast is written in more than twelve million letters. The project engaged 600 scientists from Europe, North America, and Japan, directed by André Goffeau from the Catholic University of Louvain in Belgium. Sequencing methods evolved as the project developed, but it took a decade to complete at an estimated cost of $40 million. The price has come down since then. In 2013, investigators from Germany and the United States sequenced the genomes of more than 700 yeast strains at a cost of around $50 per genome.[8] The sequencing time was reduced to two weeks. The latest methods promise complete sequences within a few days.

When I met my wife, Diana, in 1991, she was a doctoral student in biochemistry, sequencing genes the old way, digesting DNA into short fragments, and labeling the letters using radioactive phosphorus-32. The radiolabeled phosphorus came in little lead containers, called pigs, which was used to tag DNA fragments separated on gels and the sequences—AAATGCGCATGCCA—were deciphered from ladder patterns of dark bands on X-ray film. The whole procedure was immensely time-consuming, but it changed the way of biology. This classical dideoxy sequencing method was adapted for the yeast genome project by introducing fluorescent labels that lit up the DNA fragments under ultraviolet light, which was a lot safer than using radionuclides, and by running the gels in sequencers that scanned the barcodes automatically. Without this automation, it would have taken a century to read all of the letters in the yeast genome.[9]

Getting the sequence was a major advance in the research effort to understand the relationship between the genes inside a yeast cell and its structure and function. This exploration of yeast genetics had begun in the 1930s with the pioneering studies of Øjvind Winge. Winge was a Danish investigator who worked at the Carlsberg Laboratory in Copenhagen, established by the famous Carlsberg Brewery.[10] Winge designed a method for carrying out specific mating reactions between yeast strains using their ascospores, and demonstrated that the fungus behaved according to Gregor Mendel's rules of inheritance. These experiments added to the growing sense that *Saccharomyces* could be the perfect experimental organism for genetic research.

In the 1940s, Carl and Gertrude Lindegren discovered the **a** and α mating types of yeast. The Lindegrens' lab at Washington University in St. Louis enjoyed funding from the Anheuser-Busch Company, whose products include the iconic Budweiser lager. (For obvious reasons, major breweries have always supported yeast research.) Using their knowledge of the yeast mating system, the Lindegrens went on to create the first chromosome map for *Saccharomyces*, which showed the relative positions of genes on chromosomes.[11]

Carl Lindegren was a complicated man. Seemingly at odds with his groundbreaking contributions to genetics, he refused to accept that DNA carried genetic information. In a display of remarkable obstinacy, he persisted in this belief even after the structure of DNA was announced in 1953.[12] He also favored ideas about acquired inheritance proposed by the Soviet geneticist Trofin Lysenko, who rejected Mendelian genetics. Setting these bizarre ideas aside, his contributions to yeast research, in collaboration with his wife, served as one of the foundations of the later genome project.

The 6,000 genes that code for yeast proteins occupy around three-fourths of the total sequence of the twelve million letters.[13] The identity of the majority of the proteins encoded by these genes has been established. They are involved in everything the cell does, from importing sugars through the cell membrane, to metabolizing these sugars in the cytoplasm, making vesicles and cell wall materials, and repairing damaged DNA. More than 200 proteins spelled out by the DNA are signaling molecules that allow the cell to respond to environmental changes and control specific processes in different regions of the cell. Combinations of proteins determine the lifespan of the cell (1,603 proteins), its sexual behavior (1,228 proteins), the formation of buds (403 proteins), and the choreography of its death (124 proteins). The functions of proteins overlap, so that many of the products of individual genes have a hand in lots of mechanisms.

Analysis of the genome begins with the search for open reading frames, or ORFs, which often correspond to genes. ORFs are strings of letters in DNA that begin with the instruction to START the transcription process and end with the instructions to STOP it. Transcription is the formation of a messenger RNA (mRNA) copy of the DNA gene, and this mRNA molecule is translated into protein on ribosomes. DNA → RNA → protein is how information is liberated from genes. DNA sequences are written in three-letter instructions called codons. Most meaningful codons specify the individual amino acids that are strung together to form proteins. The instructions to start and stop transcription of the ORF are also types of codon: ATG is the single codon that says START, and TAG is one of three codons that says STOP. When these are found in the genome, the sequences of codons between START and STOP are prime candidates for genes that encode proteins. These sequences can be compared with the DNA of other organisms archived in huge genetic databases. In this fashion, researchers begin to annotate the genome, linking genes to their protein products.

Once a gene that encodes a protein has been found, the next step is to understand the function of the protein. This is straightforward when genes with similar sequences are found in multiple species. Genes that encode versions of proteins that are widespread in nature are likely to share a common evolutionary origin. These are called homologous genes. Genes that encode the hexokinase proteins that break down glucose are examples of homologous genes that are quite easy to spot. Although the sequences of hexokinase genes vary a lot between species, all of them incorporate a common subset of As, Ts, Gs, and Cs, called a sequence motif. Sequence motifs are signposts for genetic researchers.

The roles of other genes are more difficult to identify from their sequences, and the best way to get at their functions is to study the behavior of mutant strains of yeast in which the standard sequences have been disrupted. Natural mutations are caused by errors that occur during the DNA replication process which is associated with budding. In the more delicate yeast strains, one of these changes in the genome happens each time the cell divides, but the rate is considerably lower in robust types. Mutant strains are produced in the laboratory by treating cells with chemical mutagens or by exposing them to ultraviolet light. These treatments damage DNA structure, producing mutations in a random fashion throughout the genome. Greater precision is achieved by introducing mutant copies of specific genes into yeast chromosomes. This is called "site-directed mutagenesis."

Tetrad analysis is a classical technique used by geneticists to determine whether or not a particular strain of yeast is carrying a mutation.[14] It was developed in Winge's lab at the Carlsberg Institute and was adapted by the Lindegrens for chromosome mapping. Tetrad refers to the quartet of ascospores produced by crossing cells of the two mating types. For genetic analysis, the four spores are separated to different spots on a culture plate using glass dissection needles controlled with a micromanipulator. Between ten and twenty tetrads are processed in this way on a single culture plate, producing forty to eighty spots, each occupied by a single spore. The growth patterns that develop when the plates are incubated provide information on the nature of the mutant genes. For example, two of the four cells from a tetrad may fail to form colonies because they carry a mutation that interferes with budding. This encourages further experiments to determine how the normal gene operates. Skilled investigators can separate up to sixty tetrads in an hour, but this is exhausting work. In recent years, the method has been simplified with the automation of some of the steps.

Researchers are also working on a hands-free technique to sort cells using an instrument called a flow cytometer that reads genetic barcodes in the tetrads.[15]

Studies on yeast strains carrying mutations have been the mainstay of *Saccharomyces* research for decades and are the basis for much of our understanding of gene function. After the unveiling of the genome, investigators launched an ambitious—and ominous-sounding—Yeast Deletion Project to engineer a complete set of thousands of strains, each lacking one of the 6,000 genes.[16] The method involves replacing each natural gene with an engineered DNA sequence that contains a gene, called *KanMX*. This modification destroys the original gene and enables researchers to confirm that they have created the correct mutant, because *KanMX* confers resistance to a lethal antibiotic called geneticin. Yeast cells that grow in the presence of geneticin must be mutants. Complete sets of these mutant strains are stored in freezers in Europe and the United States.

One of the surprises of the deletion experiments is the finding that most of the individual genes can be disabled without killing the fungus. Just one in five genes was found to be essential for survival when the yeast was grown on a plentiful supply of sugars. This apparent redundancy is partly explained by the overlap between the functions of genes that allow for workarounds when the normal genome is compromised. The ancient genome duplication event, described in Chapter 1, provided *Saccharomyces* with the huge informational buffer of a copy for every gene.[17] The majority of these copies have been lost since this informational overload, but even after 100 million years of editing as many as one in ten yeast genes have backups.

The other reason for the apparent superfluity of genes is that many of these instructions for making proteins are needed only when the fungus is growing in less than ideal circumstances. Living on a culture plate, bathed in sugars, in a warm lab incubator is a pretty cushy existence. Consider how much more your body is doing when it is running through the woods, dehydrated and hungry, and pursued by wolves, versus soaking in a hot tub with a glass of champagne. Things are a lot tougher in nature, for man and yeast, and both of our genomes were sculpted to meet the exigencies of an alfresco existence. This exploration of yeast genetics as a way to understand the workings of Homo sapiens is the lure that funds much of the basic research on the sugar fungus.

Each protein encoded in the genome does not work in isolation, and efforts are underway to look at the effects of deleting genes two at a time. Interactions between gene products are a key to understanding how cells function. A method called two-hybrid screening is used to reveal which proteins are forming pairs as the fungus goes about its business.[18] Successful reactions are indicated by the expression of reporter genes that are introduced into the yeast cell to serve as flags for successful protein-to-protein hook-ups. One of the popular reporter genes allows the yeast to grow on a special nutrient medium. Another turns colonies blue when protein pairs are formed. These reporters are activated when a molecule called a transcription factor binds to the DNA. This transcription factor has two components and only works when they are brought together. By attaching different proteins to the complementary parts of the transcription factor, growth or a color change indicates that a particular protein pair has become associated. This is the first step in figuring out pairwise interactions between proteins.

Computer automation of two-hybrid screens allows investigators to examine the interplay between thousands of genes and construct maps that show tens of thousands of interactions between gene pairs. Synthetic genetic array analysis (SGA) is another method used to study these genetic interactions in cells.[19] It relies on a robotic instrument to manipulate colonies of mutant yeasts. These robots are fitted with clusters of pins, like miniature beds of nails, which pick up tiny dollops of yeast and inoculate multiwell plates or agar slabs. SGA probes tens of thousands of combinations of mutant yeast strains to reveal interactions between genes. The SGA robot made by Singer Instruments in the United Kingdom comes "with a beer bottle opener, fitted as standard," on the outside of the housing, reflecting the fictitious level of recreation made possible by this ingenious workhorse of the yeast lab.[20]

The results of these remarkable experiments are illustrated with wiring diagrams in which colored spots indicate genes, genes are clustered according to common function, and thin lines connect genes that work together.[21] In the resulting web of interactions, genes that participate in the same cellular process, like protein synthesis, stand out as foci among the myriad dots and lines that radiate out to the more distant genes with which they are associated. These genetic landscapes are reminiscent of astronomical maps in which stars are linked to highlight constellations.

With so much scientific scrutiny using an armamentarium of powerful genetic techniques, we might expect that the job of annotating the yeast genome—figuring out what each of the 6,000 genes does—is almost complete. But twenty years after the publication of the genome, the function of one in ten of yeast's genes remains unknown.[22] The function of these orphan genes are mystifying because they do not resemble DNA sequences from any other species. Equally puzzling are ORFs that look like functional genes but do not appear to code for anything useful at all.

About one-fourth of the yeast genome has been identified as noncoding DNA, meaning that it is not used for manufacturing proteins.[23] We know that a small portion of this slice of the genome is dedicated to making RNA that is not read into proteins, and that other parts of the noncoding sequence can hop around the genome and assume various roles in development. The mobile sequences are called transposable elements. But most of the noncoding DNA appears to be useless. These sequences may qualify as junk DNA. Having surplus sequences represents a burden for the cell because it has to be copied, along with the important part of the genome, every time a bud is produced. All twelve million letters have to be assembled and strung together even if they have no informational value, which seems to be a waste of energy. Worse still, the junk is not arranged conveniently, but is scattered between the coding sequences of the genes and as interruptions within the genes.

Think about the inconvenience of printing nonsense of this kind in a dictionary. (Richard Dawkins is the master of wielding encyclopedias and dictionaries as metaphors in his elucidations of genes.) The 11th edition of the *Pocket Oxford English Dictionary* is the right length for this illustration.[24] With about 1,000 pages of words and definitions, we would need to insert a total of 250 pages of junk to mimic the yeast genome, but not by doing so in a convenient appendix at the end of the book. In this bewildering publication, noncoding DNA sequences would be represented as nonsensical words between meaningful words, and as nonsense within individual definitions. Searching for the definition of solipsism, which we can use to represent a single gene, we find the preceding entry, soliloquy, but have to flip through a page or two of nonsensical words and letters before we find the word we are looking for. Once we find it, solipsism is defined as, "the view *rabbits have silky ears* that the self is all *sea cucumber* that can be *tofu vindaloo* known to exist."

The yeast genome looks like this bizarre dictionary, but the fungus manages to sidestep the rubbish without a hiccup, never trying to read the entries between the functional genes into proteins. Noncoding DNA sequences stuck inside genes demand special attention. These are called "introns." They are transcribed into RNA, then cut out, or spliced, before the uninterrupted proteins are made on the ribosomes. It is as if we read the definition of solipsism once, including the *tofu vindaloo*, then cut the italics to make sense of the entry. The genetic mechanism is called splicing and is one of the dazzling cellular workarounds that have evolved to cope with the clutter in the genome.

The situation for humans appears to be much worse, because only 2 percent of our larger genome is read into proteins. We have around 19,000 genes and are overwhelmed with junk DNA. If we printed the three billion letters of the human genome in the format of the *Pocket OED*, the publication would run to 1,000 volumes and occupy fifty meters of library shelving. Yet only twenty volumes would be required to cover the instructions for making proteins. This needle-in-the-haystack construction of genomes is one of the multiple and irrefutable proofs of unintelligent design. Some molecular biologists believe that junk DNA is brimming with useful information that we have not learned how to read yet. Others say that junk is junk.[25] Favoring the junk-is-junk conclusion is the observation that onions have five times more DNA than us.[26] This is termed the "onion test." Either a lot of onion DNA is rubbish, or it takes more DNA to make an onion than a human. The least humbling conclusion is that onions have accumulated a lot of genetic baggage in the form of no-longer-coding DNA on their way to becoming onions. Ranked against onions, the human genome is a comparatively uncluttered instruction manual, and yeast's blueprint is the paragon of lucidity.

Noncoding DNA is mostly flotsam picked up along the unbroken thread of life from the origin of the first replicating cells to today. It comes from copies of genes that were replicated and then suffered mutations that rendered them useless. It comes from infections by viruses that inserted their genes into chromosomes under the misapprehension that they might sit there for a while, get copied along with the rest of the yeast genome, and reawaken in the form of future generations of viruses. Our DNA is full of these sorts of fossil viral infections. Junk DNA is like the totality of fathomless memories, some that jog the conscious mind once in a while, others that reappear in senseless dreams—"Why am I dressed as a ballerina and preparing to go on stage with Mick Jagger?"—the quotidian refuse of life. We put it aside, get on with the tasks of the present, and do our best to ignore the interference from random remembrance. Likewise, the cell keeps copying the junk and ignores most of it when it makes proteins.

The reason that this genetic refuse has not been discarded must be that it is easier to bring it along for the ride than to remove it. Unless sequences of noncoding DNA interfere with the expression of neighboring functional genes, there is little pressure to remove them from the genome. These sequences may be invisible to natural selection, causing no harm or good. Another reason that these masses of DNA are retained in genomes is that they may serve as a repository of sequences that can be called upon in the evolution of new genes. This might explain where some of the orphan genes in yeast come from. Eighty years after Øjvind Winge's exploration of *Saccharomyces* as a model organism for biological research, with our record of spectacular breakthroughs in yeast genetics, we are left with a sense of having scratched the surface of a greater inscrutability. And there is something awesome in owning our innocence, knowing that there is so much more to learn.

The relative simplicity of yeast is championed as one of its virtues as an experimental model. This is also a limitation when we consider the bridge between the unicellular fungus and multicellular human. The wild life of the yeast is often spent in the company of clones carrying identical genes, which means that each cell is of equivalent importance to every other individual in the surrounding soup. The same clonal identity holds for the trillions of human cells in our bodies, but our cells work as a collective. We have nerve cells and bone cells and retinal cells, and all thirty trillion of them participate in the goal of getting our genes into the future. Only sperm and egg cells transmit the genes, but they carry the same genes as all of our other cells. Yeast has none of this division of labor found in animals and plants. It is far simpler too than the related fungi that produce mushrooms. The cells in the stalk of an edible morel act as a platform for the indented head of the mushroom from which spores are discharged. The airborne spores released from the morel fruit body are the only cells that carry the genes of the fungus into the future.

With no distribution of tasks among different cell types, every yeast cell has the potential to act like a sperm or egg by merging with a cell of the opposite mating type and combining resources to form ascospores. Sexual reproduction is one of the more sophisticated characteristics of yeast that set it apart from bacteria, which do not bother with sex. The single yeast cell has two more elements in its developmental repertoire that allow it to extend itself beyond the loneliness of the single cell. Some strains of *Saccharomyces* produce filaments when they are starved of nitrogen.[27] These grow when the daughter cell does not separate from its mother, and goes on to produce a daughter of her own. Strings of yeast cells develop in this fashion and the elongation of each of the cells creates a growth form that resembles a chain of sausage links. This is an expression of clonal cooperation, still a long way from reflecting any maternal support, division of labor, or self-sacrifice.

The advantage to the starving cell in switching from budding to filamentous growth is mechanical. Budding cells pile up on surfaces forming glistening globs; filaments allow the fungus to explore solid food materials by burrowing into their surroundings. Once upon a lifetime, incidentally (for the benefit of my biographers), I studied the mechanics of invasive growth in fungi for fifteen years. This process allows pathogenic species to penetrate the living tissues of plants and animals.

After mating and making filaments, the third bravura performance by yeast is called "snowflaking."[28] The fungus has to be encouraged to do this in the lab and the result serves as a powerful demonstration of evolution, quicker than breeding domestic animals, but every bit as wondrous in its power. When yeast is growing in grape must, or another sweet liquid, its cells tend to separate after budding rather than forming clumps. Clumping can be a liability for a microbe growing in liquid because it increases the weight of the particles, causing them to settle toward the bottom of the vessel and lose out on access to sugars dispersed in the fluid above. Yeast is driven to clump in the snowflaking trials by gently centrifuging the culture so that any cells that stick together tend to spin out to the bottom of the culture tube. By transferring cells from the bottom of the tube into fresh medium, we can select for clumping ability, and after sixty transfers the population of yeasts is dominated by snowflake patterns of conjoined cells. As snow-flakes get larger, the probability of creating unique patterns increases because buds do not form at precisely the same place on every dividing cell. No pair of yeast snowflakes, like their name-sakes of crystallized water, is quite the same.

The liability of sinking in nature is transformed into a charac-teristic that is crucial for survival of the yeast population in the lab. This is an example of artificial selection, akin to the breeding of Great Danes by picking the tallest dogs from successive litters of hunting dogs.[29] Darwin began his 1859 masterpiece with a description of the practice and consequences of artificial selec-tion, explaining how the deliberate choice of individual farm ani-mals and racing pigeons was employed by breeders to cultivate particular characteristics. *Artificial* selection by humans serves as a simpler model for *natural* selection in which changes in environ-mental conditions affect the frequency of different versions of genes in populations of species.

Snowflake yeast offers a bulletproof illustration of this process. Mutations in the yeast DNA that favor a long-term connection between mother and daughter cells increase the weight of the par-ticles in the growth medium and are carried with cells to the bot-tom of their tubes when they are centrifuged. Because these cells are transferred to the fresh broth, the mutations will be propa-gated. The fact that snowflake yeast loses the benefit of floating throughout the fluid is immaterial. Its evolution is driven toward this unnatural growth form because it has no freedom to do otherwise in the lab. Snowflake mutants are sheltered from the blast furnace of natural selection.

Investigators have identified the precise mutations in yeast responsible for snowflaking by comparing the genomes of the normal yeast with the snowflake form. These genetic alterations appear in a gene called ACE2, which is a transcription factor that manages the expression of genes involved in cell separation. Research on snowflake yeast is significant for evolutionary biologists interested in the origins of multicellularity.[30] Snowflake yeasts may exist somewhere in nature, but the separation of daughter cells is the normal behavior for the fungus. Mutations in ACE2 occur all the time, but clumping cells are disadvantaged and the mutation is lost from populations when it arises. Saccharomyces has been entirely successful as a single cell for millions of years.

Yeast does not help, at least directly, in the search for the origins of our multicellularity. Instead, we have to retreat along the branches of human ancestry, past relatives of the treeshrew with a penchant for palm nectar, to reptilian and fishy predecessors, to the sponges. Sponges, along with comb jellies and a curious creature called a placozoan (or flat animal), are the marine organisms that we find in the foundations of animal evolution. Simpler still are the choanoflagellates (Figure 12), or collar flagellates, whose cells use a single tail or flagellum to create fluid currents to draw in their bacterial food.[31] The bacteria are strained from the water by a ring of tiny finger like projections that form the collar to which their name refers. Each cell resembles a shuttlecock. Cells of choanoflagellates that form colonies cling together using protein junctions, creating multicelled structures with greater order than a yeast snowflake.

Genetic comparisons between these simplest of animals and the fungi show that we were born from the same ancestors. We are much more closely related to yeast and mushrooms than we are to plants, slime molds, seaweeds, and the greater diversity of life unseen. This is recognized by the classification of fungi and

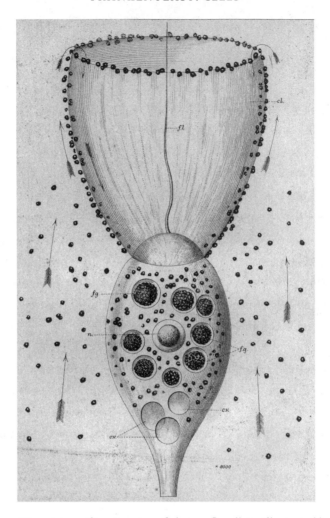

Fig. 12. *Monosiga gracilis*, a species of choanoflagellate, illustrated by William Saville Kent. The drawing shows a cell that has been supplied with particles to study the feeding mechanism used by this microorganism. When bacteria are trapped on the collar of the cell, they are absorbed into food vacuoles within the cytoplasm.

animals in a supergroup called the Opisthokonta. For the poor creationist vexed by her kinship with chimpanzees, this demonstration of fungal affinity must be terrifying. Our split from the fungi occurred around one billion years ago, deep in the Precambrian when all life lived as single cells. One billion years is a big chunk of time, 20 percent of earth's history, or 10 percent of the life of our sun before it runs out of hydrogen. And yet, we should recognize something of ourselves when we look at yeast under a microscope. The essence of the yeast and human cell is the same. Revisiting the classic diagram of cell structure discussed at the beginning of this chapter—the scattering of burrito-shaped mitochondria, slice of Golgi apparatus, and so on—we might expect the semblance between yeasts and humans to dissolve beyond this generalization. Yet we remain unified by equivalences that run much deeper.

Recalling our earlier inventory of the genetic riches that constitute *Saccharomyces*, we saw that only one in five of its 6,000 genes are essential when the fungus is privy to unlimited food, kept warm, and protected from stress. The sequences of around half of these indispensable genes are sufficiently similar to instructions in the human genome that they must have evolved from the same ancestral DNA sequences. Matching genes like these are called orthologs, and many of them are interchangeable between species. This has been tested by replacing yeast genes with human genes.[32] To do this, copies of a human gene are installed in circular DNA molecules called plasmids and the plasmids are transferred into yeast. The type of plasmid used in these experiments can operate like extra chromosomes in the yeast cell, serving as a novel source of one or more foreign proteins. Introduction of the plasmid into a mutant yeast strain in which the corresponding fungal gene can be turned off reveals whether the human gene works as a substitute.

Investigators found that 176 of 414 essential orthologs in yeast could be complemented with human genes. In some cases, the human genes worked when the positions of only 10 percent of their As, Ts, Gs, and Cs matched the yeast sequences. Replaceability was determined by gene function. Where yeast genes were dedicated to making and breaking down lipids, amino acids, and sugars, human orthologs worked very well. Yeast fared poorly, however, when the investigators tried to replace genes involved in the replication and repair of DNA, or genes crucial for directing cell growth.

Another approach to re-engineering yeast is to fabricate new versions of the fungus containing synthetic chromosomes. The natural version of *Saccharomyces cerevisiae*, Sc1.0, was fashioned after the providential genome duplication error made by an ancestral yeast in the Cretaceous. One hundred million years later, Sc2.0 is being built in the lab.[33] The artificial version of the first of yeast's sixteen chromosomes, SynIII, was completed in 2014, and another five came off the molecular assembly line in 2017. SynIII is a sleek version of chromosome III that lacks junk DNA and has been augmented with special sequences called loxPsym sites. loxPsym sequences are used to switch the positions of chunks of DNA along the chromosome. The other five chromosomes have the same basic architecture. This ability to scramble the genome in a controlled fashion—to change the position of genes and introduce foreign genes that encode proteins outside yeast's natural stockpile—will allow researchers to develop new strains of yeast with novel properties. This form of accelerated evolution will also advance the goal of creating a yeast strain with a minimal genome, a servile *Überpilz* or superfungus. Sc2.0 will be a semi-artificial form of life that will operate according to the desires of its engineers.

Sc2.0 will be a bioengineered cell, a mixture of nature and artifice, a "replicant" for those familiar with the 1982 movie *Blade Runner* and other science fictions. If the genome of Sc2.0 is capable of operating on its own, without any interference from the genome of Sc1.0, the fungus will make its own machinery for protein synthesis, replicate and repair its DNA, formulate a system of internal membranes, make buds, and even reproduce if we engineer Sc2.0**a** and Sc2.0*α* strains. The mitochondrial power plants in the cell present a challenge because they have their own small genomes. Engineers may work with the natural mitochondria of yeast, or create new ones with their own synthetic minigenomes.

Ethical concerns are muted by the fact that genetic engineers are modifying yeast rather than human cells, but there are issues of biosafety that must be addressed. The project leaders argue that the parents of the strain used for this reinvention of yeast were isolated in the 1930s from rotting figs in the Central Valley of California.[34] This heritage sounds innocent enough. The strain is labeled S288C, which is the same fungus used in the original sequencing project. It grows very well in the lab, but carries a number of mutations that make it reliant on specific culture conditions and reduce its fitness in the wild. These disabilities should keep it from taking over the planet. A second weakness that will foil S288C's pursuit of planetary domination is that it does not form the filamentous chains of cells described earlier in the chapter. This means that it cannot hunt for food by invading its surroundings if it escaped from the lab.

On a more alarming note, these defects apply only to the *unmodified* yeast strain S288C, and the biotechnological promise of Sc2.0

is that we will be able to remake the fungus into anything we choose. In their "Statement of Ethics and Government," the Sc2.0 team state, "We will conduct and promote our work on Sc2.0 for the benefit of humankind."[35] Along these lines, Sc2.0 may be reworked to make antibiotics and biofuels, produce synthetic fibers, or clean up toxic waste. This is very laudable, but if the synthetic yeast escapes its sachet, a malevolent scientist might seek to re-engineer *Saccharomyces* into the bearer of lethal toxins or into an agent that causes brain infections. Like the majority of fungi, yeast does not grow in human tissues very often, but on the rare occasions when *Saccharomyces* spreads in the body it is very unpleasant. Frankenyeast boosted with virulent genes from other microbes could be very nasty indeed. "As such," the Sc2.0 researchers respond, "we are exploring the possibility of additional engineered vulnerabilities to further decrease the likelihood of viability outside of the laboratory in an effort to minimize the chance of harm, should there be an accidental release."

The prospect of creating living organisms from the foundation of species crafted by evolution is exciting and troubling. Just as Einstein's rendering of the equivalence of mass and energy in $E = mc^2$ made the atom bomb theoretically possible, Watson and Crick's solution to the structure of DNA unleashed the prospect of the recreation of life. There are sound reasons for proceeding very cautiously with this reimagining of the sugar fungus. Young yeast biologists should read, or reread, Mary Shelley's *Frankenstein* as a source of wisdom on the importance of knowing when it is time to turn back.

It is interesting that Sc2.0 is underway at a time of renewed interest in the way that the first cells evolved in the pre-biological world. Mary Shelley wrote long before the revelations of Darwinian evolution, but she was intrigued by the work of Erasmus Darwin, paternal grandfather to Charles. In *Zoonomia*, published in the 1790s, Erasmus imagined that all life had arisen from "one living filament... possessing the faculty of continuing to improve by its own inherent activity, and of delivering down those improvements by generation to its posterity."[36] The origin of the first cells remains the greatest mystery in biology.[37] By creating Sc2.0 we are adapting the existing plan that has been twisted and tweaked, revamped by mutation, duplicated and edited for tens of millions of years. In identifying what we think is junk and reorganizing the placement of instructions within a refashioned genome, we are selecting a yeast that meets our needs rather than those of the rest of nature. This work may be trivial compared with the feat of making a cell from scratch, but it is an important step toward mastering the construction process that has worked very well without us.

If Sc2.0 becomes a functional synthetic organism, we will have access to a complete gene-by-gene picture of the instructions that make it work. Supercomputers will allow us to model all of its interacting parts, from one millisecond snapshot to the next. This simulacrum will be as complex as the cell itself and exist as a virtual fungus, living and breathing until we pull the plug. We will be able to change the virtual environment and watch the beast evolve. Then, in short, humans will have recreated life. Bloggers will opine under the headline, "Playing God."

And still our comprehension of the way of the cell will remain elusive. Consider the mismatch between scientific facts and the gut-level appreciation in astronomy. The Orion Nebula is our closest site of star formation. It was recognized as something special in the seventeenth century, a few decades before Leeuwenhoek first spied yeast with his single-lens microscope. This cloud below the belt of the constellation is more than 1,300 light years from earth, or thirteen quadrillion kilometers (10^{16} km), and it is twenty-four light years across. These immense distances are well-established facts, but they are unfathomable. Traveling at 60,000 kilometers per hour, which is the speed of NASA's New Horizons spacecraft that passed Pluto in 2015, it would take twenty-five million years to reach the Orion Nebula.

In the same vein, we have mastered the tools to modify the 100 million-year-old miracle of yeast, but remain humbled by the magnified view of the pullulating globs of fungus. Even when we have learned how to program a computer to simulate every chemical reaction occurring in each moment in the life of the yeast cell, our minds will be little closer to grasping the whole beyond the metaphor of a flamenco dance.

5

The Little Yeast on the Prairie

Biotechnology

It's late July on a hot afternoon, under a cloudless blue sky, and I am cycling north along Stateline Road which straddles the border between Ohio and Indiana a few miles from my home. The high is supposed to top 93°F today, 34°C. Protected by sunscreen and sustained by periodic drafts of water, this is ideal weather for a long ride. Having grown up in England, where the summer sky was gray more often than blue, a day like this one has been one of the pleasures of immigration. Climate models suggest that opportunities for this kind of innocent recreation will be subverted by the ferocity of future Midwestern heatwaves.[1] The level of CO_2 in the atmosphere is more than eighty parts per million higher today than it was when I was born, and higher than it has been for at least 800,000 years. We could use some help—something to put us on the path toward carbon neutrality.

The road is mine today—no cars or trucks—and the asphalt is hemmed in by corn (maize) crops on both sides and blistering in places. It is getting really hot. Chirping from spur-throated grasshoppers, *Melanoplus ponderosus*, provides a soothing aural backdrop for this monocultural tour. Eastern tailed blue butterflies, *Cupido comyntas*, are a welcome sight. They flit in pairs, strays from the remnants of wild vegetation within this ocean of green. Next month this wider view will recede as the corn reaches up three meters (over nine feet) and the road will seem more like a tunnel. The corn along this stretch of the road is grown for *Saccharomyces*, which turns sugars from the yellow kernels into ethanol. According to advocates of corn ethanol, yeast is our redeemer, a miracle from the microbial world that may save humanity.

Ninety million acres of American farmland are dedicated to growing corn, *Zea mays*, and 40 percent of the harvest in 2014 went to bioethanol production.[2] This bio-agro-technological enterprise occupies more farmland than the total area of England. There were more than six million farms in America in the 1930s, before the Dust Bowl and a combination of economic pressures depopulated rural counties across the Great Plains. There are four million fewer farms today, but the average size of these agricultural holdings has more than doubled. The annual corn harvest supports more than 200 bioethanol plants with the capacity to generate fifteen billion gallons (fifty-seven billion liters) of ethanol—around one tenth of the volume of gasoline combusted in the United States.[3]

Biofuel production has come at great costs to the environment and to taxpayers. Great swathes of grassland have given way to corn and soybean. (Soybean oil is used to produce biodiesel without using yeast.) The grassland lost to these biofuel crops includes pasture and hay lands, retired cropland, and some scraps of native prairie. The rate of grassland loss in America—percentage not acreage—is comparable to the deforestation in Brazil, Malaysia, and Indonesia.[4] Financing for the construction of bioethanol plants relies upon generous government subsidies. Corn growers, agribusiness lobbyists, and senators representing Iowa—the biggest producer—and neighboring states claim that this is an essential investment to meet our energy needs. We are witnessing an agricultural revolution and *Saccharomyces* is the catalyst.

Corn ethanol is a controversial fuel, marketed as a clean, environmentally friendly solution to our energy needs by the biofuel industry, and vilified as a waste of valuable resources by its critics. The industrial virtue of corn lies in its ability to absorb water and dissolved minerals from the soil, and make masses of sugars from thin air. At first glance, the use of yeast to create a simple burnable fuel from slightly more complex sugar molecules manufactured by corn seems to be a good bet. If we illustrate bioethanol production using an arrow to show corn absorbing CO_2 from the atmosphere, another arrow pointing to yeast, and a third to ethanol + CO_2, all looks good (Figure 13). We have met our energy needs without adding any greenhouse gases to the atmosphere. But there is more to this.

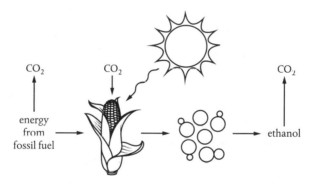

Fig. 13. The flow of energy and the absorption and release of CO_2 associated with the production and consumption of bioethanol produced from corn. Sunlight powers photosynthesis by plants that absorb CO_2 from the atmosphere and manufacture sugars. Sugar from processed corn is metabolized by yeast to produce ethanol. When the ethanol is consumed as a fuel, CO_2 is released into the atmosphere. Because CO_2 emissions from burning ethanol are balanced by the CO_2 absorbed by the plants, advocates of this technology claim that the process is carbon neutral. This picture is complicated by the consumption of oil and gas, and associated CO_2 emissions, associated with today's highly mechanized industry of cereal agriculture.

Entropy, or disorder, increases when sugar is converted into alcohol and CO_2, which means that the direction of the reactions—from sugar to alcohol—is favorable from the perspective of the energy input needed to make this work. If we reversed things and tried to use alcohol and CO_2 as the raw materials to make sugars we would need a great deal of energy to make this work. The same argument holds true for using coal as a fuel. It is easier to turn coal into greenhouses gases, than to make the black stuff from CO_2. (Although, having said this, plants succeeded in doing this 300 million years ago using photosynthesis powered by the Carboniferous sun.) The problem with coal is that burning it sends ancient carbon back into the atmosphere, producing a net increase in greenhouse gases. Bioethanol is more subtle in its shortcomings.

The plant packs these sugar molecules into its kernels, the raw material for bioethanol production. The rest of the plant is superfluous from the biofuel perspective—at least using the current industrial methods—because these tissues are made from cellulose and other big molecules that cannot be digested by yeast. After the harvest, the fibrous corn plants are usually left to decompose in the fields and the cobs lie scattered on the ground. The decaying root system also helps to stabilize the soil against erosion until the next crop is planted, and the rotting stems and leaves fertilize the soil. Some farmers allow cattle to graze the fresh stover soon after the kernels are harvested, and others combine the fibrous waste with other materials to produce fodder. There are plenty of calories in these by-products, trapped in cellulose and related polymers called hemicelluloses, waiting to be released in the multichambered digestive systems of the ruminants.

Even after harvesting the cobs, yeast is still one step from the food it needs to make alcohol. Yeast is a spoiled child of nature, one who refuses to eat anything but candy, which explains the necessity of unlocking sugars from the starch grains in the plump kernels. This is achieved using expensive enzymes purified from other microorganisms. Amylases in cereal flour are critical in bread-making, for releasing sugar from starch. Corn kernels contain their own amylases, but they do not work fast enough for the purpose of biofuel production. The solution to liberating glucose from corn is an ingenious two-step feat of microbial technology. Step 1, called "liquefaction," involves using a heat-stable version of the enzyme α-amylase to solubilize the starch. Most bioethanol plants use α-amylase produced by the bacterium *Bacillus licheniformis* (which, curiously, lives on bird feathers in the wild). Liquefaction takes place after starch slurry from the corn is combined with pressurized steam in a jet cooker. Step 2, called "saccharification," releases glucose from the cooled mash by means of glucoamylase enzymes produced by the filamentous fungus *Aspergillus niger*, or its distant relative *Trichoderma reesei*. Both species belong to the ascomycetes like *Saccharomyces*.

Fig. 14. Modern bioethanol plant.

Bioethanol is made from the glucose in fermentation tanks that would not look out of place in a brewery (Figure 14). The resulting "beer" is quite strong, with an alcohol concentration comparable to wine. Degassing to expel CO_2, and distillation, followed by a process called "molecular sieving"—to remove the remaining water—produces 200-proof absolute ethanol. Absolute ethanol is twice the strength of 100-proof vodka, which has an alcohol concentration of 50 percent. Mixed with gasoline, the distillate becomes E85 ethanol, a fuel blend with a maximum ethanol concentration of 85 percent that powers flex-fuel vehicles.

To assess the costs and benefits of corn bioethanol properly, we have to consider the energy used to cultivate corn. Corn is an expensive crop, with conventional fuels needed to power the machinery, liberal applications of fertilizers, herbicides to keep the weeds away, insecticides to kill the pests, and lots and lots of water. Potassium fertilizers are produced from minerals extracted from deep mines and nitrogen fertilizers are made by the Haber process—an intensive chemical method which is responsible for 1–2 percent of total global energy consumption. Insecticides and herbicides are petroleum products. Most of the corn grown in the Midwest is transgenic, expressing the gene from the bacterium *Bacillus thuringiensis* (Bt) that encodes the Bt toxin that destroys caterpillars of the European corn borer. Unfortunately, planting Bt corn does not eliminate the use of conventional pesticides sprayed on the crop.

Many of these chemicals are transported thousands of miles to be spread and sprayed on Midwestern farmland. Runoff from soils bubbles in shallow ditches around the fields and trickles into larger streams that unite to form tributaries that feed into the Ohio River. The Ohio joins the Mississippi in Cairo, Illinois. Farther south, the waste fertilizers from the farmland spread through the southern bayous, and plankton blooms exhaust the oxygen in the infamous dead zone of the Gulf of Mexico. A foreshadow of the threat to marine life is obvious on my bicycle rides in Indiana, where fat tangles of bright green algae clog the streams draining the cornfields. Only the oldest residents can remember a time when there were any fish in these waters.

Ethane produced from natural gas in Pennsylvania is piped beneath my bike route and it may be joined by a bioethanol pipeline before too long. Most bioethanol is transported by rail and truck today, because it causes corrosion cracking that damages pipelines. Ethanol is very flammable too, but the technology is advancing, and a 2,700-kilometer route for a bioethanol pipeline has been proposed. This would complete the conversion of the Midwest from rural farmland into a vast biorefinery.

Modern industrial farming is an exercise in rolling big boulders up hills without recouping enough energy as the boulders run downhill. The energy input is greater than the energy output. This is the second law of thermodynamics and nobody, not even the rain, can disobey the second law. Critics of biofuels seem to have won the scientific argument for now, but only if we do not consider the alternatives. Continuing to meet our energy needs with power plants that run on coal, oil, and natural gas is worse than using bioethanol, albeit marginally, in terms of greenhouse gases.[5]

A similarly massive energy input is need to grow sugarcane, but at least the resulting crop provides sucrose, rather than cornstarch, which yeast can ferment as soon as it is squeezed from the plant. Sugarcane has been cultivated in Brazil for 500 years and bioethanol production from the crop became a serious business during the oil crisis in the 1970s. Brazil is the second-largest bioethanol

Fig. 15. Sugarcane field in Brazil.

producer today, with a yield of 6–7 billion gallons of fuel from sugarcane grown on 80,000 square kilometers of farmland in 2014 (Figure 15). The state of São Paulo in south-eastern Brazil is the country's largest producer, making it the agricultural facsimile of Iowa. Most of the Brazilian crop is cultivated on abandoned cattle pasture, but before the herds arrived this land was covered with hardwood forests that teemed with life. (We are so familiar with the relatively modern phenomenon of wide-scale deforestation in South America that it is easy to forget that North American agriculture is founded on an earlier biological and humanitarian holocaust. Before the European migration in the nineteenth century, Iowa was dominated by 116,000 square kilometers of tallgrass prairie that was occupied by Indian tribes including the Ioway or Báxoie, Sauk, and Meskwaki or Fox.[6])

The cane part of sugarcane refers to the stick formed by the stems of a number of species of grasses in the genus *Saccharus*.

Sugarcane belongs to a taxonomic category called a "tribe" that sits within the larger family of grasses. Corn is a member of the same tribe, along with sorghum, which is another plant tapped by the biofuels industry. The whole cane of sugarcane is consumed in the service of ethanol production, but the leaves stripped from the plant during harvesting are usually left to rot in the fields, like corn stover. The sugar released from the chopped and shredded canes is sucrose, or table sugar, the disaccharide made from one glucose molecule and one fructose. Yeast deals with sucrose from the canes in the same way it works through wheat dough—by breaking the sucrose molecule into two parts and consuming the component monosaccharides. The mass of fibrous residue from the processed stalks, called "bagasse," is used to fire boilers that furnish heat and generate power for the bioethanol plant. Fermentation and distillation methods are similar to those used to generate corn ethanol, but there is no need to use amylases to break down starch.

Pure gasoline (E0) is no longer sold in Brazil, and newer cars run on flex-fuel blends ranging from 25 percent ethanol in E25, to pure ethanol in E100 fuel. E27 is the standard blend sold in Brazilian gas stations at the time of writing. Governmental standards for fuel blends are adjusted regularly because the cost of ethanol varies according to annual sugarcane yields. So while supply and demand and a farrago of geopolitical issues control oil prices, Brazilian ethanol is more responsive to a spell of bad weather and its effect on the sugarcane harvest.

The yeast strains used in Brazil in sugarcane fermentation are wilder microbes than those used to ferment glucose in North American bioethanol plants. When sugarcane processors added commercial yeast strains that had been used in baking to their fermenters, they found that these bread yeasts were pushed aside by local environmental strains of *Saccharomyces* as the fermentation progressed.[7] These interlopers were more competitive than the commercial strains and out-budded them into oblivion. Once in control, however, the wild yeasts often made a mess of the process by generating foam and failing to match the ethanol yield of the slower-growing strains coveted by bakers.

In the 1990s, bioethanol researchers became excited when they found a novel yeast strain that combined the aggression of the wild yeasts with the high ethanol yield of the fungi recruited from bakeries. This strain, designated "PE-2," has become the darling of the Brazilian industry.[8] PE-2 is an example of a diploid yeast strain that has two sets of chromosomes. Diploid strains come from sexual mergers that do not result in the birth of four ascospores. By forgoing spore formation, the cells remain bigger than the haploid strains with a single set of chromosomes. They replicate all thirty-two chromosomes with each cell cycle and can bud as diploid cells indefinitely. The reason for the vigor of diploid yeasts is not an obvious consequence of having a double helping of DNA in each cell, but they do tend to have a higher level of ethanol tolerance than the haploid strains from which they originate. Diploid yeast strains are now commonplace in the global biofuel industry.

Bioethanol production from sugarcane is a punishing experience for yeast. Tons of yeast cells are added to the fermentation vats to accelerate the development of a dense suspension of the budding cells. The fermentation process runs continuously for 6–8 months per year, with two or three runs per day until the seasonal supply of fresh-cut sugarcane is exhausted. At the end of each run, when the liquor is harvested, the yeast is collected by centrifugation, washed in sulfuric acid for two hours, and returned to the fermenter with a fresh supply of sugar for its next 6–10-hour shift.

The acid wash kills bacteria that build up as each fermentation proceeds. Lactic acid bacteria are the main malefactors in bioethanol fermenters, just as they are the spoilers in winemaking.[9] The taste of bioethanol is irrelevant, but the production of lactic acid and acetic acid by the bacteria dampens ethanol synthesis by the yeast, which is germane. Yeast does not thrive in sulfuric acid, but refreshed by the chemical scrub it is ready to get back to work.

In addition to the shock of the acid bath, the yeast cells are dehydrated by the concentrated sugar at the beginning of each fermentation, stressed by the heat in the fermenter, and come close to being poisoned by their own alcohol. We would lose our skins in sulfuric acid and, once flayed, would shrivel in the sugar and pickle in hot ethanol. The trials of the climbing boys indentured to chimneysweeps in Victorian Britain come closest, perhaps, to the microscopic servitude of sugarcane yeast.[10] These comparisons are bereft of scientific meaning, but may encourage greater respect for the tribulations of *Saccharomyces* on our behalf.

Great efforts have been made to develop yeast strains with higher tolerance to heat and alcohol. One of the approaches has been to adopt the artificial selection strategy used to "evolve" snowflake yeast colonies (that we saw in Chapter 4). For thermotolerance, a flask is inoculated with yeast and incubated at 40°C, rather than its preferred temperature of 30°C.[11] Most cells quit growing and die, but the survivors are transferred to a fresh flask kept at the elevated temperature. When this punishment is repeated daily for three months, the yeasts emerge with new fortitude and grow twice as fast at 40°C than their ancestors. This is a significant change in physiology: 40°C is the same temperature as a hot bath. The evolutionary adaptation that makes this possible is a mutation in a gene called *ERG3*, which controls the synthesis of lipid molecules that sit in the cell membrane. Membrane damage is one of the consequences of temperature stress in yeast, and changes in *ERG3* activity that result in the production of a sterol, called fecosterol, seem to underlie the performance of the yeast subjected to artificial selection.

The speed and power of this selective routine may be used to produce future yeast strains with all kinds of useful biotechnological attributes. Now that we know that the *ERG3* gene is a good target for creating industrial yeasts that can cope with high temperatures, we can replace the native version of the gene in any yeast strain. Alterations in membrane chemistry can also be featured in Sc2.0, Frankenyeast, but without the simple experiments with warmed flasks we would not know which gene to mutate. Experimental evolution can tell us where to look.

Another approach to improving the performance of yeast in the bioethanol industry is to adjust its growing conditions. This does not involve any alteration in the genetics of yeast and is, in a sense, the kind of thing that brewers have been doing unconsciously since the Sumerians figured out how to make beer from cereal grain. In the biofuel industry, the alcohol tolerance of the fungus determines how much ethanol can be produced before yeast pickles itself. High levels of ethanol damage the cell membrane by making it leaky, and treatments that allow it to keep working as a gatekeeper between the cell and its surroundings allow the yeast to withstand more alcohol. Experiments show that the addition of potassium ions (K^+) to the nutrient broth used to grow yeast, coupled with a modest reduction in acidity, enable the cell to maintain a healthy balance of ions in the cytoplasm as the ethanol concentration rises.[12]

The explanation is quite complicated, but seems to hinge on the way that the change in chemical environs makes it easier for transport proteins in the yeast membrane to keep working despite the harm done by the ethanol. This is a temporary fix because the damage to the membrane does not go away. Nevertheless, this environmental manipulation may be useful in keeping things going toward the end of a fermentation run when the ethanol levels become critical. The fact that potassium transport is a critical player in ethanol tolerance also suggests that genetic enhancements to existing transport proteins in membranes may be profitable.

The use of a powerful genetic editing technique called CRISPR/Cas9 promises a new era of bioengineering in which yeast will be manipulated in ways unimagined by earlier biologists.[13] Longstanding approaches to gene editing require the targeted deletion of genes followed by several steps to incorporate a replacement that confers new characteristics upon the resulting mutant. These techniques have been highly successful, as we have seen with the humanization of the yeast genome discussed in Chapter 4, but CRISPR technology saves an enormous amount of time by chopping out the target genes and substituting their replacements in one go.

CRISPR has been described as "a molecular Swiss army knife," which allows multiple edits to the yeast genome in a single sweep.[14] The method relies on using an RNA molecule as a guide for an enzyme that does the cut and paste job. The enzyme was discovered in bacteria, which use the CRISPR mechanism to destroy viruses that infect their cells. It evolved as a kind of primitive immune system that protects bacteria from viruses that inject their DNA into their cells. The method can be used to make changes to single letters in a DNA sequence.[15] This level of precision is very exciting for clinical researchers because it offers the possibility of correcting human mutations that cause many devastating illnesses.

CRISPR raises plenty of ethical concerns too, whether the method is used to re-engineer a microbial workforce or, as many worry, ourselves.[16]

The essential problem lies with the fact that there are more than seven billion of us, more consumers than ever. We are here in great numbers, in universal pursuit of creature comforts, and in desperate need of a scheme for global cooling. Few solutions to reducing carbon emissions seem as practical as enhancing our partnership with *Saccharomyces*. If yeast could be reengineered to adopt more of the character of a filamentous fungus without losing its yeastiness we might end our reliance on fossil fuels forever. Corn and sugarcane would be superfluous too, because this saintly mutant could feed on compost and wastepaper.

Filamentous fungi that rot wood use enzymes called cellulases to break down cellulose, transforming fallen trees into pulp that dissolves into the soil. Many of the species that are best at this miraculous act of recycling are mushrooms, including fungi that shed their spores from giant shelves that fruit from the rotting wood. The artist's conk, *Ganoderma applanatum*, is a wood rotter with global distribution. (The fungus gets its common name from the practice of etching the delicate underside of the fruit body with a stylus, whose tracks leave a fine stain in the desired design.) It begins life as a spore that germinates on a wounded tree, one of trillions of microscopic particles that leave the narrow tubes beneath the fruit bodies of their parents and waft on the forest breeze. Germination produces the first thread, or hypha, of an intricate colony, or mycelium, that eats its way into the wood, extracting sugars from the cellulose and other complex molecules that constitute the hard tissues that supported the tree throughout its long life.[17] Cellulase enzymes are released from the tips of the growing filaments and chop the cellulose polymers up into glucose monomers, which are burned for energy within the mycelium. This is called "white rot."

If wood chips, agricultural waste, recycled paper, and discarded packaging are shredded, soaked in water, and enriched with some nitrogen, all of them can be fed to white rot fungi. White rot fungi that liberate glucose from these cellulosic materials require oxygen to complete the digestive process. Using this efficient method of respiration, woodland fungi tap the maximum number of calories from every sugar molecule they absorb. This is a fundamentally different strategy from the seemingly wasteful growth of yeast, with or without oxygen, which leaves a lot of the calories behind in the form of ethanol. Can we combine the toolkit of the white rot fungus with the speedy growth characteristics of yeast? This is an ambitious mission for fungal biology, but it may be our best hope for leaving fossil fuels in the ground and putting the brakes on climate change.

Significant advances have been made using agricultural waste, rather than corn kernels or sucrose from sugarcane, as the raw material or feedstock for bioethanol production. This technology relies upon the use of purified enzymes to release sugars from wheat straw and other fibrous plant matter before *Saccharomyces* gets to work. Ethanol produced in this fashion is referred to as a "second-generation" biofuel. The technical challenge lies in the presence of lignin, which shadows the cellulose in plant tissues. Lignin is a complicated molecule of branching chains of aromatic rings that white rot fungi like the artist's conk have to remove before they can digest the cellulose. A special group of fungi called "brown rotters" have evolved a mysterious process that allows them to get at the cellulose without having to break apart the lignin. This biochemical sorcery is evident from the rendering of fallen trees into brown cubes that crumble to the touch. The brown color comes from the lignin left behind by the brown rot fungus. If we understood how this worked we might have another plan for transforming yeast.

Using the present technology, cellulose and other molecules in the raw feedstock are treated with concentrated acids or purified enzymes to release sugars for fermentation by yeast. The use of corrosive chemicals presents safety issues, but the process is very efficient, converting up to 90 percent of the cellulose, and other polymers, into fermentable sugars. GM strains of the ascomycete *Trichoderma reesei* are the main source of cellulases in the biofuel industry. Using enzymes to break down cellulose into sugars is much safer than using concentrated sulfuric acid or hydrochloric acid, but purified enzymes are very expensive.[18] Proponents of these second-generation fuels claim that the technology is carbon-neutral; critics argue that the process generates more CO_2 than oil and gas.[19] The truth hinges on the detailed analysis of every part of the production chain, and a simple resolution to the debate is impossible.

Attempts at modifying *Saccharomyces* to do the entire job of cellulose fermentation have met with modest success. Yeast strains have been transformed with one or more genes from bacteria that encode cellulases, conferring the ability to digest cellulose.[20] This is a dramatic makeover for the sugar fungus, like turning a lion into a vegan. Levels of ethanol production are encouraging when the GM yeast is fed purified cellulose, but drop when cellulose comes in the more complex form of pulp made from rice straw. Other carbohydrates, called hemicelluloses, are combined with the cellulose and lignin in plant tissues. These are an alternative source of fermentable sugars, whose breakdown requires another class of enzymes called hemicellulases. The genetic transformation of yeast to grow on these substances is another research front in the biofuels industry.[21] It made our beer, raised our bread, and yeast may keep earth habitable and humans mobile if we can persuade it to become an omnivore.

In addition to its Herculean efforts in ethanol production, yeast has found gainful employment as a producer of butanol, isobutanol, and biodiesel. Isobutanol—$(CH_3)_2CHCH_2OH$—is attracting the greatest interest among investors because it packs more energy per liter than ethanol. This means that isobutanol could be blended with kerosene and other petroleum products to design greener jet fuels.[22] Isobutanol has been described as an alcohol that acts like a hydrocarbon—the distinction being that alcohols contain hydroxyl groups (–OH), and hydrocarbons, refined from oil, are chains of pure carbon and hydrogen. In addition to its potential value in jet fuel, isobutanol beats ethanol as a gasoline additive for cars, and is less volatile and corrosive than ethanol, which simplifies transportation. Yeast is somewhat reticent about making these more advanced biofuels, preferring to get back to ethanol synthesis when it is given plenty of sugar. Efforts are being made to curb these natural tendencies through genetic modifications that include the amputation of some of its native DNA and augmentation with genes from bacteria. We probe the genome of the sugar fungus, searching for hidden strengths and weaknesses, for genes we may remove and others we might improve upon, to teach yeast new magic in the service of man.

Biofuels are the most obvious fruits of recent biotechnological meddlings with yeast, but there have been many successes in other areas of human concern. On summer bicycle rides, the heat becomes most evident when one breaks to take a swig of water and listen to the crickets. This is an opportunity for mosquitos to avail themselves of a sip of English blood. Indiana was a malarial hellhole until the Great Black Swamp and other wetlands were drained in the nineteenth century. The worst of the mosquito-borne illnesses in this part of the world is West Nile virus, which is, thankfully, very rare in the Midwest. Mosquito bites are irritating, of course, and insect repellant is a necessary preparation for a ride, run, woodland hike, city stroll, canoe trip, or snooze in a garden chair. For those bitten by malarial mosquitos, GM yeast may become a lifesaver.

Artemisinin is an antimalarial drug extracted from sweet wormwood, *Artemisia annua,* a plant used in traditional Chinese medicine. Tu Youyou, who discovered and developed the drug, was one of three scientists awarded with a Nobel Prize in 2015 for their contributions to the treatment of parasitic infections.[23] She was the first Chinese scientist to receive a Nobel in the category of Physiology or Medicine. Artemisinin kills the malarial parasite, *Plasmodium falciparum,* and the drug has saved millions of lives. Developing yeast as an artemisinin factory promises to reduce the cost of the drug by providing competition with suppliers of the plant extract and ensuring a continuous supply of this vital medicine in the developing world.

The technology for artemisinin synthesis is complicated, and its development was funded by a major grant from the Bill and Melinda Gates Foundation.[24] Work began in 2004 and the delivery of the winning yeast strain was announced at the University of California, Berkeley in 2012. This magical fungus tenders a hybrid biochemical pathway that combines the use of naturally occurring yeast enzymes with enzymes encoded by genes appropriated from the wormwood plant. The artemisinic acid molecule released from the yeast cells has to be rearranged a bit to produce the final drug, but the use of a single strain of the fungus to generate the precursor of the life-saving medicine is considered a major success in the new science of metabolic engineering.

This financial support from the Gates Foundation was critical because malaria is a disease of the developing world, concentrated in Africa, where healthcare spending is limited and profits for pharmaceutical companies are rather slim. Drug development is a costly business, supported by companies with little incentive to sink resources into drugs that patients cannot afford. This modest declaration of capitalism lays bare the mendacious nature of drug companies that profess enthusiasm for wiping out tropical diseases, and agribusinesses that offer fantastical advertisements about their commitment to feeding the world's poor. In the absence of profit, few companies are in a position to give a fig about malnutrition. Rare exceptions to this rule are manifest when a business wagers on the public relations benefit of a full-page advertisement showing sub-Saharan children beaming with relief at the prospect of their imminent emancipation from some dread riverine parasite.[25]

Diabetes is a more profitable illness than malaria, and engineered strains of *Saccharomyces* have been chugging out insulin since the 1980s. Insulin is a protein that is assembled from two parts—an A-chain and a B-chain—encoded by a single human gene, *INS*. Expression of a modified version of this gene in yeast releases an inactive form of the protein from the cell, and this is treated with an enzyme called trypsin to produce the active hormone. The global insulin market in 2015 was valued at more than $20 billion, and with a growing number of insulin-dependent patients, it is expected to double in value by 2020.[26] Danish company

Novo Nordisk is the leader in manufacturing insulin using yeast. Yeast produces half of the insulin used to treat diabetes; the other half comes from the gut bacterium E. *coli*. Eli Lilly and Company, based in Indianapolis, uses the bacterium to manufacture its best-selling insulin product called Humalog. Until the 1980s, Eli Lilly extracted insulin from piles of pig pancreas liberated from the poor animals raised in barns the size of aircraft hangers through-out the Midwest. The bacterial process is cheaper than using yeast, but E. *coli* retains the insulin inside the cells, making it necessary to bust the cell wall apart to get at the drug. The more sophisticated packaging and processing machinery inside the *Saccharomyces* cell allows yeast to release the insulin into the broth of the fermenter. This makes purifying the drug a doddle.

Yeast is a cell factory for many other pharmaceutical agents, including blood products and vaccines. Vaccination against human papillomavirus (HPV), which is the main cause of cervical cancer, has become routine using virus-like particles synthesized by yeast. The shell of the active virus is assembled from seventy-two copies of a single protein called L1. Incorporation of the viral gene encoding L1 in yeast allows the fungus to produce masses of the protein units that associate spontaneously to form complete shells or capsids. The resulting particles are virus-like, rather than completely viral, because they are empty of the viral DNA necessary to spread the infection. Injection of the recombinant protein provokes antibody production and protects the patient against HPV exposure. Gardasil, which is the trade name of the HPV vaccine, has been a good investment for Merck and Company.[27] In the six years since the introduction of the vaccine, the prevalence of viral infection has fallen by 64 percent in teenage women between the ages of fourteen and nineteen, and by 34 percent in twenty to twenty-four year olds.[28] The consequences of vaccination will be evident, as time passes, in an anticipated reduction in the number of cases of cervical cancer.

Ocriplasmin is another engineered yeast product, designed for older patients with an eye problem called symptomatic vitreo-macular adhesion. This is produced by yeast transformed with a human gene that encodes a protein-digesting enzyme. The enzyme is injected into the eye to dissolve connections between the vitreous humor and the retina that distort vision with the natural deformation of the vitreous humor associated with aging.[29]

Insulin and ocriplasmin are pretty simple drugs whose production in yeast is achieved via relatively straightforward manipulations of the fungal genome with individual genes from humans. Artemisinin production belongs to a different category of advanced biotechnology involving the re-engineering of entire metabolic pathways. With the encouraging breakthrough of CRISPR technology, it seems feasible that continuing modifications of *Saccharomyces* will enable us to employ the single-celled fungus as a GM surrogate for the unnatural manufacture of a treasure trove of valuable natural products. This is exciting and terrifying.

For all the good that may be done, there is a very troubling side to yeast and its potential for manipulation. In the past twenty years, signs of stress in rural Midwestern communities are obvious from the increasing number of shuttered businesses, absence of young people in small towns, and sentiments about the imminent arrival of Christ on church signs. My favorite sign has a yeasty resonance: "The best vitamin for a Christian is B1."[30] There is more immediacy to another genre of signs along the straight roads through the cornfields: "Entrance Needle Exchange Program," "Indiana State Police, Meth Tip Line, Call 1–800-***-****," and "Drug Checkpoint

Ahead." Heroin, cocaine, and methamphetamine use has sky-rocketed in small-town America and addiction to opioid pain-killers is rampant.[31] The relationship of this social tragedy to the sugar fungus comes in reformatting yeast to home-brew heroin.

Heroin, or diacetylmorphine, is made by boiling morphine with acetic anhydride. This adds a pair of acetate groups ($COCH_3$) to the ringed morphine molecule that is purified from dried latex harvested from opium poppies. Acetylation makes heroin more soluble in fat than morphine, easing its passage from the bloodstream into the brain, causing immediate euphoria and, in the slightly longer term, its antonyms.

The latex that exudes from the scored seed heads of the poppy, as well as the whole plant minus the seeds, is used by the pharmaceutical industry to extract a range of natural opiates, including codeine and thebaine. Codeine is a relatively mild painkiller, good for some types of headache and effective as a cough suppressant. Thebaine has the opposite effect of morphine, acting as a stimulant rather than a narcotic, and is the precursor of oxycodone, marketed as OxyContin in the United States. Oxycodone is prescribed as a more powerful analgesic, or painkiller, than codeine. The problem with this medicine is that it can be highly addictive and, in some cases, cause as much misery as heroin.

Wild *Saccharomyces* does not make any of these compounds. If it had acquired the trick of making opiates in its evolutionary past, palm wine drinkers would have bathed their nerve endings in such tranquilizing spirits that early *Homo sapiens* would have sleep-walked itself into the fossil record.[32] The biochemical pathway for making opiates is considerably more complicated than alcoholic fermentation. It involves eighteen steps, each catalyzed by a separate enzyme, compared to the twelve-step program for converting glucose to ethanol and CO_2. On its own, yeast produces the amino acid tyrosine, which is the substance that opium poppies use as the basis for morphine. Using this as a starting point, genetic engineers have transformed yeast with a gene that encodes an enzyme used by sugar beets to make L-DOPA from tyrosine.[33] L-DOPA is the brain chemical that is converted into dopamine and other neurotransmitters—molecules that transmit nerve impulses from nerve to nerve, and nerve to muscle or gland cell. Subsequent steps in opiate synthesis have also been incorporated into yeast with one significant exception—one that is keeping the genie in the bottle.[34] Once the missing enzyme is introduced we will have cobbled together a yeast strain that makes thebaine, codeine, and morphine. This fungus would be the idol of Big Pharma, and, once liberated from the lab, ruin the market for opium poppies. More importantly, the zombification of communities stemming from home-brewed heroin would make today's drug epidemic seem like a children's birthday party.

Heroin made by yeast could, perversely, increase the market for a riot control spray that is also made by yeast. This high-tech chemical mixture was developed by an Israeli company that advertises the product as, "Non-toxic. Non-lethal. No nonsense."[35] The "100 percent eco-friendly" formulation is secret, but yeast conducts the fermentation that generates a blend of compounds whose smell has been described as "An overpowering mix of rotting meat, old socks...topped off with the pungent waft of an open sewer."[36] Adding to the irony of using a spray produced by yeast to disperse people destabilized by a drug made by yeast is the name of the spray—"Skunk," which is one of the slang terms for heroin, as well as marijuana. Assuming the yeast spray resembles the appalling smell of a striped skunk, *Mephitis mephitis*, it seems well suited for getting people to move where you want them to.

Everywhere we look, my evident yeast obsession notwithstanding, we find the enterprises of yeast. Back on my bicycle voyaging through the ocean of GM corn, it is the absence of wildlife that is so striking. One of the comeliest animals in the farmland is the inch-long argiope spider, *Argiope aurantia*, a black and yellow arachnid that strings its web at the field margins with a striking zigzag of silk called a stabilimentum. Twenty years ago I saw these spiders all the time; now I stop riding to pay homage when I spot one. It is the totality of industrialized agricultural practices that damage the rhythms of wildlife.[37] There is too little flying around to be captured in their zigzag webs. Loss of the natural wizards of web construction seems tragic at a time when transgenic yeast is being used to make spider silk.[38] A company in California called Bolt Threads is growing yeast fitted with spider genes in fermenters and extracting fibers of synthetic silk for use in garment manufacture.[39]

Unlike their counterparts in the study of human genetics, yeast scientists avoid the charge of hubris, playing God, and yet the environmental consequences of this biotechnological insurgency are potentially greater than any advances in manipulating our genomes. Biofuels made by yeast may help to keep the planet habitable, especially if we can refashion the fungus to consume fibrous agricultural waste. If this experiment fails we may go the way of the dinosaurs. Abandoned by its industrial partner, *Saccharomyces* will vanish from the shining steel fermenters, retreat to its ancient floral haunts, and bud forever, just like it did before we came along.

6

~

Yeasts of the Wild

Yeast Diversity

Microbes that cause lethal infections hold a privileged position in human concerns. Outbreaks of the Ebola virus are the stuff of global news alerts, and the development of antibiotic-resistant bacteria worries everyone that gives this some thought. The celebrity status of the sugar fungus is a singular case in our relations with the microbial world, of a species known for the good things that it does for us. In light of the centrality of *Saccharomyces* in our affairs, it is not surprising that people are unaware that this is one of many hundreds of yeast species. Few of these other yeasts have any direct interactions with humans, but their significance in our lives is phenomenal. The majority of them feed as decomposers, breaking down biological waste and recycling nutrients. Without these ecosystem functions, none of us would be here. Civilization has always been predicated on these activities, which makes the story of these microbes a crucial part of this book.

Evolutionary biologists describe yeasts as polyphyletic, which means that they have evolved from a number of different ancestral groups rather than a single common ancestor.[1] Yeasts contrast with monophyletic groups, like the primates, which have sprung from a common ancestor. (Incidentally, the common ancestor of the primates resembled the pen-tailed treeshrew.) The business of being a yeast comes down to existing as an ovoid fungal cell that reproduces by forming buds. This behavior has developed and spread independently across several distinct limbs on the fungal side of the tree of life. Most of these species are of the ascomycete kind, related to morels and truffles, but others dot a separate branch, one they share with gilled mushrooms. A third grouping is allied with pin molds that bloom black on melon rinds and rotting tomatoes. We find these microbes wherever we care to look.

Benson is an English village at the confluence of the River Thames and a chalk stream, which flows from a spring that bubbles in a neighboring village. A battle fought there in 779 is mentioned in the ninth-century *Anglo-Saxon Chronicle*, using the name Benesingtun, in Old English, which became Bensington, subsequently Benson. The name of the village is pertinent to this book, because it was Latinized in 1986 to describe a new species of yeast called *Bensingtonia ciliata*.[2] The fungus looks a lot like *Saccharomyces* when it is budding, but the resemblance ends with this microscope snapshot of maternal industry. *Bensingtonia* is more closely related to rust fungi that attack wheat crops than it is to *Saccharomyces*, and shares nothing in lifestyle with the sugar fungus.

Bensingtonia was brought to light by the famous mycologist, Cecil Terence Ingold, who moved to Benson upon retirement in the 1970s and continued his researches on fungi in a home laboratory for the next twenty years.[3] Ingold found his yeast growing on the fleshy tissues of a jelly fungus called the wood ear mushroom, *Auricularia auricula-judae*. The wood ear is a brown jelly fungus that is thought to be a modern representative of the earliest kind of mushrooms that flourished at the end of the Carboniferous Period. It is a mycological coelacanth, a living fossil.[4]

Bensingtonia grows on the slithery surface of the wood ear, eking out a living by consuming the mucilage that gives this mushroom its gelatinous texture, and dispersing itself via airborne spores. The mechanism of self-propelled flight used by Ingold's yeast is a marvel of natural engineering, one of the wonders of the microbial world. *Bensingtonia* creates an escape module in the form of a bud on a short spike that points above the surface on which the yeast colony is growing. A second or two before the launch, a tiny bead of water expands from the bottom of this bud, or spore, like a dewdrop, then makes contact with the side of the spore and collapses on its surface. This movement causes an almost instantaneous shift in mass that powers a short jump into the air, high enough for the spore to be swept away in the slightest breeze.

When a colony of *Bensingtonia* is viewed with a microscope, the spores and their water drops are visible, fleetingly, but disappear almost as soon as they are brought into focus. The surface of the colony appears to be blinking, drop...drop gone, another drop... drop gone...*ping, ping, ping* across the surface of the culture. If the yeast is cultivated in a recognizable pattern by drawing the inoculation loop across the agar in the shape of a cross or a letter, the mirror image of the yeast cells is reproduced on the underside of the lid of the culture dish (Figure 16). The little fungus spray-paints its ceiling with spores. This behavior is indicated in the name "mirror yeast" for species, like *Bensingtonia ciliata*, which possess this catapult mechanism for dispersing spores.[5]

By curious coincidence, I spent my childhood in Benson, living at the other end of the street from the house occupied by C. T. Ingold and his wife, Nora. He merged, as far as my adolescent brain was concerned, with the other elderly villagers, a person of no consequence to me. The strange part of this tale is that I knew

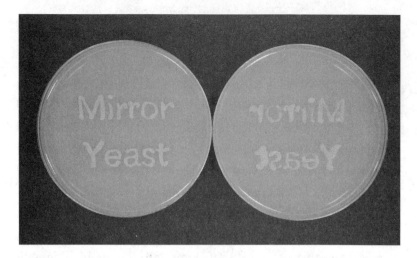

Fig. 16. Mirror image of the salmon-pink colony of the mirror yeast *Sporobolomyces* made by placing the open Petri dish of an active culture inoculated with a fine paint brush (left) on top of a Petri dish containing fresh agar culture medium (right). The culture on the left discharges its spores from the surface of the agar. The spores land on the agar below and multiply by budding to give rise to a second culture that mirrors the overhanging text.

nothing of Ingold's profession until after I had begun work as an intern in a mycology lab at university. My awakening came when I chanced upon one of his books and saw from the back cover that the author lived in Benson, Oxfordshire. We became friends after this revelation, walking in the woods when I visited my parents, the retired professor becoming a mentor. Logic would suggest that this relationship fostered my highly esoteric scientific interests, but I had set on them at least a year before we met. Imagine a teenager who leaves home and becomes a professional tightrope walker, then learns that his childhood neighbor—in a rural village with less than 5,000 residents—was a world famous funambulist and you will grasp the improbability of this facet of my biography.[6]

Another personality deserves mention in this part of the yeast story. Arthur Henry Reginald Buller (1874–1944) was obsessed with mirror yeasts. He studied a fungus called *Sporobolomyces roseus*, named for its rose red coloration, exploring its flight mechanism in exquisite detail. Like *Bensingtonia*, *Sporobolomyces* spores use drop movement to jump into the air. He also studied the budding process in the fungus, and tracked the division and movement of the nuclei as each daughter cell developed. He expressed his dedication to his fungus in a poem, "The Sporobolomycetologist," which he set to a musical score and performed at a Christmas party. One stanza will suffice:

> Perhaps in heaven, where angels are,
> His yeast thoughts will persist:
> He was an ardent Spor-o-bol-
> O-my-cet-o-log-ist![7]

A founding member of the science faculty at the University of Manitoba in Winnipeg, Buller was an extraordinary scientist, advancing the study of fungi like no other mycologist in history, driving the field from a largely descriptive endeavor concerned with collecting and naming species into the era of rigorous experimentation. His work on *Sporobolomyces* represented a small part of his immense contribution to mycology. In 1930, he was honored with the naming of *Bullera*, a new genus of mirror yeasts. His response, perhaps unsurprisingly, was poetic, resulting in a brief ditty, which begins:

> O Bullera, yeast-genus named for me,
> Thou'rt wan and from all trace of pigment free;[8]

"Uncle Reggie," as his students called him, was aware that most of his poetry was doggerel, and loved the real thing. As he was dying from a brain tumor, he consoled himself with reading John Milton, frustrated that he could not complete another volume of his magnum opus, a book series titled *Researches on Fungi*.[9]

Long before the family tree of the mirror yeasts was explored by DNA analysis, Buller's observation of water drop formation on their spores indicated that they were allied with the basidiomycete fungi. This was inferred from the fact that the catapult mechanism is a unique feature of the basidiomycete group that includes mushrooms, shelf fungi, jelly fungi, and plant-killing parasites called rusts and smuts. Buller never saw the spores in the act of jumping, because they move so fast—faster than fleas. One moment the spores are motionless, a drop of water clinging to their surface, the next they have disappeared from the microscope field of view. The motion of the drops on to the spore surface and the coinciding launch, was not revealed until a century later when the motion was slowed with high-speed video cameras running at 100,000 frames per second.[10]

Jumping spores are limited to the yeasts that are counted as distant relatives of mushrooms. There is nothing like this gymnastic display among the ascomycete relations of *Saccharomyces*, although some have evolved other kinds of spectacular spore launches on this branch of the fungal tree too. Species of yeast called *Metschnikowia*—named after Élie Metchnikoff, the Russian-born researcher who identified the yeast in the 1880s—produce spores that look like bee stings with backward-pointing spines (Figure 17).[11] The spores are fifty times longer than the budding cells of this fungus, and housing them requires the growth of an extension that works like a torpedo tube. These needle-shaped spores are ascospores, produced after sex, like the rounded cells that develop after mating between the **a**- and *α*-cells of the sugar fungus.

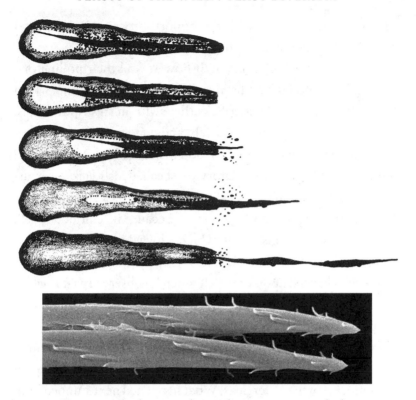

Fig. 17. Ascospore discharge in the predacious yeast *Metschnikowia*. Drawings show the end of the ascus dissolving in preparation for ascospore release (compare first and second image), followed by the forcible ejection of the pair of needle-shaped spores. The barbed tips of the spores are evident when they are viewed with a scanning electron microscope.

Metschnikowia yeasts found in morning glory flowers use these darts to infect the little beetles that pollinate them. For a fungus that cannot get airborne on its own like a mirror yeast, hitching a ride inside an insect is a great strategy—as we have seen with the sugar fungus. Yeasts growing in the nectar are slurped up by the beetles and are shot from their tubes when they reach the beetle gut, perforate the lining, and hold fast with their barbs.[12]

This provides the yeast with the opportunity to reproduce and shed daughter cells in every flower visited by the insect.

Besides their associations with flowers and their pollinators, *Metschnikowia* species have been isolated from brine shrimp, which live in salty lakes, and water fleas that swim in freshwater lakes, ponds, streams, rivers, water barrels, and puddles.[13] The water flea, *Daphnia magna*, which is used as food in fish tanks, is plagued by a ferocious species of the predatory yeast called *Metschnikowia bicuspidata*. Metchnikoff recognized that *Daphnia* was a superb organism for studying infection processes because its chitinous shell, or exoskeleton, is transparent, which allowed him to watch what happened inside the animal, as it was happening, with a microscope. He chanced upon the yeast when he observed that *Daphnia* fished from an aquarium were filled with "a massive accumulation of fungal cells."[14]

Looking more closely, Metchnikoff saw the yeast spores puncture the intestinal wall and slip into the surrounding body cavity or hemocoel of the water fleas. What happened next stunned him. The crustacean blood cells moved toward the invading spores, glommed on to their surface, and swallowed them, bit by bit, with fragments of the needles becoming absorbed into their own substance. Metchnikoff had discovered phagocytosis, one of the fundamental defense responses of innate or cellular immunity. He described his work in an 1884 paper titled, "Über eine Sprosspilz-krankheit der Daphnien: Beitrag zur Lehre über den Kampf der Phagocyten gegen Krankheitserreger," which I quote because *Sprosspilzkrankheit* and *Krankheitserreger* convey so much more intensity than their English translations of *yeast-like fungus* and *pathogen*. And Metchnikoff was an intense individual. He attempted suicide first with morphine, then tried again by injecting himself with the bacterium that causes relapsing fever, and yet survived to pursue a scientific career punctuated with vitriolic disagreements with his peers.[15]

For his work on phagocytosis, Metchnikoff was awarded the Nobel Prize in 1908, along with Paul Ehrlich, who had found the adaptive arm of the immune system that relies on antibodies rather than phagocytes.[16] Interest in the infection of *Daphnia* by *Metschnikowia* continues today, with the disease serving as an experimental model for evolutionary biologists. The Red Queen hypothesis, developed in the 1970s, refers to the arms race between predators and their prey in which the increasing determination of the hunter drives the evolution of better defenses in their quarry. To survive, hosts need to keep adapting. The Red Queen expressed this challenging circumstance very clearly in Lewis Carroll's *Through the Looking-Glass,* when she tells Alice, "Now *here,* you see, it takes all the running *you* can do, to keep in the same place."[17] Whether you are predator or prey, failure to adapt equals hunger and extinction. Life is trapped in this wheel of horror.

Metschnikowia spreads through a population of *Daphnia* like wildfire, killing the planktonic animals in the prime of their filter-feeding lives. Interestingly, well-fed *Daphnia* die faster than hungry animals when they are stricken with yeast.[18] *Metschnikowia* goes berserk when it punctures the gut of a fat water flea. Surrounded by the droplets of stored lipids, the fungus buds furiously and fills the body cavity of its host. When a sickened *Daphnia* literally bursts open, it releases tens of thousands of its needle-shaped spores into the water. Neighboring animals do not stand a chance. *Metschnikowia* hits plankton like a barrel bomb.

Resistance to the yeast infection relies on the mechanism of phagocytosis, but there is a limit to the effectiveness of these defenses. It may be less costly in evolutionary terms for the water fleas to play the odds and wait out an epidemic. *Daphnia* reproduces asexually part of the time, so that many of the individuals in a population carry exactly the same genome. Therefore, while death by yeast infection destroys the individual, the water flea genes may be transmitted by millions of other clones that escape the fungus. This means that it can make more sense for the animals to invest in feeding, growing, and reproducing, rather than evolving more elaborate immune defenses. In this fashion, the surviving water fleas can reboot their population after the yeasts crash for want of sufficient prey.

The predator plays its cards carefully too. If *Metschnikowia* is too aggressive, it can limit the opportunities for survival by killing all the available hosts before being able to spread to other parts of the lake. Dispersal is essential because this is the only way to locate fresh congregations of water fleas, swimming around with gay abandon, innocent of the approaching inferno. The most successful strategy is to plague one's hosts without reducing their numbers too precipitously.

Metschnikowia is one of many yeasts that have a predatory side to their personalities. Hidden killer instincts are revealed when other yeast species are grown in pairs in culture experiments. In the most obvious confrontations, one of the strains will stick to the other, force feeding tubes through their cell walls, and suck out the juices.[19] Species of *Saccharomycopsis* and *Candida* that live in the sticky sap that oozes from tree wounds are particularly aggressive, and can decimate entire populations of the sugar fungus when they are grown together. Far from being hapless victims of these mycopredatory yeasts, some strains of *Saccharomyces* keep their competitors at bay by releasing toxins. These poisons work by damaging the cell walls of the enemy species, making their membranes leaky, inhibiting DNA synthesis, and stopping bud formation.[20] Killer yeast toxins come from three sources. Some are encoded in genes carried by viruses that spread from cell to cell during mating. Others are carried on plasmids (little loops of DNA that carry information), and poisons of a third class are specified by genes situated in the yeast's own chromosomes.

The viral system is very complicated, with one type of virus transmitting the toxin gene and a second virus working as a helper that ensures that the toxin is made and secreted from the yeast cell. To prevent the self-destruction of the killer yeast, the formation and release of the toxin confers immunity upon the hosts to any of the poison that is backwashed after it has been released. Without this shield, the host yeast cells would be killed as they spat out the toxin and the onboard viruses would be non-starters. It all has to work like a Swiss watch. The resulting cooperative venture is a win-win for both of the viruses and their yeasts: the viruses get to spread through populations of yeast and the yeasts benefit from the destruction of their competitors. All that matters is the transmission of genes through space and time, according to the pitiless logic of nature.

Toxin release depends upon the ubiquitous cellular mechanisms of protein packaging and secretion, and the elucidation of the killer yeast story has helped scientists tease apart some of the fundamental workings of cells. The repeated evolution of these chemical weapons in viruses and within the yeast cells themselves, as well as the process of physical attack with feeding tubes, illustrates the perpetual battle between yeasts wherever mixed populations of the fungi are growing. In uncontrolled fermentations of palm sap and grape juice, the success of particular yeasts is based upon their ability to cope with changing environmental conditions. The alcohol tolerance of *Saccharomyces cerevisiae* makes this fungus the champion of most natural brews, but the expression of viral toxins provide privileged strains with an additional edge over their competitors. Researchers in the food and beverage industry show understandable interest in killer yeasts.[21] The slightest competitive edge identified in nature could make all the difference. Vintners and brewers, like their yeasts, have a lot riding on the outcome of internecine warfare among fungi. Additional interest in yeast toxins comes from the ongoing search for novel antifungal agents to treat human infections.

Metschnikowia, with its gut harpoons, ranks as an eye-catching yeast among the thousands of species that do nothing more than multiply by budding and form little blobs within larger blobs after mating. Even *Metschnikowia* is a drab speck of a fungus until the barbs on the tips of its *Daphnia* darts are magnified with an electron microscope. The smallness and the sameness of yeasts is one of the reasons that they have received relatively little attention from classical mycologists. There are so many kinds of filamentous fungi with much more interesting shapes and developmental processes that draw one's attention with even the crudest microscope. On the other hand, consider *Dipodascus*, the most ostentatious performer in the carnival of yeasts, a fungus to thrill the imagination of the weariest microbiologist.

Dipodascus was my first mycological love. We were introduced by my most engaging professor, Mike Madelin, who was interested in the fungi that grow in the creamy trails left behind slime molds.[22] (Aren't we all?) Slime molds are more closely related to amoebas than fungi. They form shimmering films called plasmodia that migrate over rotting wood and leave mucus in their wake.

A distinctive microcosmos is born in this unpromising habitat, as it is in every morsel of food on earth. Wherever we look, we find life, and wherever we find life there are yeasts. Madelin described a new species that he called *Dipodascus macrosporus*, and gave me the opportunity to study how it reproduced. As an undergraduate chomping at the bit to do real research, this seemed like the myco-logical equivalent of being offered the position of ship's naturalist on HMS *Beagle*.[23]

Mating in *Dipodascus* is not like mating in other yeasts. After two cells make contact, one of them acts like a sperm cell, trans-ferring its nucleus into the other cell, where fertilization takes place. Subsequent development in *Dipodascus* is marvelous, with the elongation of the fertilized cell into a spike that sticks up into the air. Up to thirty-two spores are formed in this cylindrical ascus and each is fitted with a gelatinous coat (Figure 18). Climax, as it were, occurs when the tip of the spike bursts and the spores ooze upwards and collect in a ball. The ruptured tip or collar of the ascus is narrower than the spores, so their jelly coats are squeezed as they make their exit. Pressure in the fluid contents of the ascus forces the spores through the collar. The formation of the exposed ball of sticky spores may aid dispersal by insects that can carry the next generation of the fungus to fresh grounds where passing slime molds leave their trails.

Spore release from the bottle-shaped asci of a related genus of yeasts, called *Dipodascopsis*, presents a mechanical problem. The bean-shaped spores in these fungi lack capsules and block the opening if a pair is pushed into the opening at the same time. Spores that turn sideways can also block the birth canal. Dried beans in a bottle get stuck like this too, until the bottle is shaken so that they drop into the neck one at a time. *Dipodascopsis* has solved

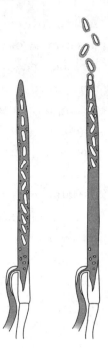

Fig. 18. Exudation of spores from the elongated ascus of the yeast *Dipodascus*.

the problem by equipping its spores with interlocking ridges, converting them into microscopic gear wheels. As the spores advance toward the opening, they rotate against one another so that their lengthwise alignment is maintained and they screw out of the ascus one at a time. The gear-wheel parallel is strengthened by the finding that the fungus lubricates its spores with an oily substance.[24]

Budding cells of *Dipodascus* and *Dipodascopsis* stick together rather than separating like *Saccharomyces* in fermenting beer, so that they produce groups of cells that resemble potted cacti with their interconnected pads. They also switch from this budding growth form to produce the elongated cells or hyphae characteristic of filamentous fungi. Many yeasts can stop growing like yeasts when they find it useful.[25] This radical shift in the developmental process is called dimorphism. It is a critical part of the behavior of many yeasts, because the formation of hyphae permits them to escape suspension in fluids and penetrate solid surfaces. This makeover allows yeasts that cause human disease to invade the walls of blood vessels or other tissue barriers (see Chapter 7).

The developmental plasticity of the fungi was recognized around Darwin's time, by brothers Charles and Louis-René Tulasne, devout Catholics who dedicated their research to the glory of God. The illustrations of microscopic fungi published in their three-volume masterpiece, *Selecta Fungorum Carpologia* (1861–5), are some of the most beautiful illustrations in the history of science (Figure 19).[26] In this sensational work, they demonstrated that single species of fungi could grow in two or more distinct forms as they progressed through their life cycles. This revelation upset the existing taxonomic order by proving that some individual fungi had been described twice and given separate Latin names.

Mycology continues to be haunted by this problem, and taxonomists are working hard to expunge duplicate names from the record. Some of the superstars in the field of fungal taxonomy—it is difficult to imagine a more obscure list of celebrities—made these errors because they were unable to grasp the degree to which fungi can change their stripes. People do not turn into birds or fish, but the varied life cycles of a fungus can compete with those fantastical metamorphoses in the paintings of Hieronymus Bosch. Even greater problems in the practice of naming fungi have come about because mycologists have sought to classify fungi according to rules established for plants and animals in the eighteenth century. Giving a fungus a Latin name or a code number has the virtue of helping different researchers ensure that they are talking

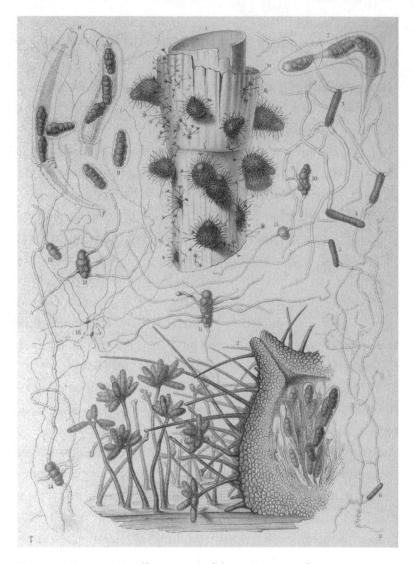

Fig. 19. Mesmerizing illustration of the ascomycete fungus *Pyrenophora polytricha* by Charles Tulasne, the Audubon of fungi.

about the same microorganism. But just because someone christens a fungus does not mean that this name refers to a distinct species.[27]

Naming problems apply to all fungi, but the sameness in the appearance of so many species that grow as single cells rather than forming colonies has made them a particularly thorny group for taxonomists. The introduction of molecular methods has been a godsend for these experts, because DNA sequences are very effective at discriminating between yeasts that look identical under the microscope. This does not mean that the naming problem can be settled easily, however, because there is no reliable measure of the genetic difference between a pair of yeasts that demonstrates that they are separate species rather than different strains of the same thing. Investigators often refer to a 3 percent difference in the order of As, Ts, Gs, and Cs in particular DNA sequences, called internal transcribed spacers (ITSs), as an indication that they are dealing with separate species. Mycologists recognize that there are lots of limitations to this adoption of a magic percentage, and yeast researchers are particularly rigorous in using multiple criteria to distinguish between species.[28] The assortment of tests includes separating, or "melting," the two strands of the DNA double helix from different yeast cultures and measuring how well they recombine. This is called DNA hybridization: efficient binding indicates genetic kinship, and loose mergers occur in dissimilar matchups. Nutritional requirements tested in culture experiments are another source of information for identifying species.[29]

Molecular genetic technology has changed the science of yeast ecology as much as it has affected yeast taxonomy. We have known for many decades that yeasts grow everywhere we bother to look, but the identification of yeast DNA in samples of soil, water, plant tissues, and the gut contents of animals has revealed the astonishing reach and diversity of these fungi. Using so-called metagenomic methods, researchers can amplify the DNA from all of the organisms bobbing around in a bottle of river water and pick out the sequences belonging to yeasts by looking for matches in an online database.

An estimated one quadrillion (10^{15}) yeast cells live in earth's rivers, 10^{17} grow in lakes, and one sextillion (10^{21}) bathe in the sea.[30] Leaf surfaces are another popular habitat for yeasts and may support as many budding cells as the oceans. Earth is a very yeasty planet. Across the globe there are common yeast species that are happy with a range of environmental conditions, and rarer fungi with more refined tastes. *Debaryomyces hansenii* is one of the generalist yeasts. It grows in soil, on leaves, and in marine habitats, and has been isolated from cheese, sausages, and (settling the question of what not to eat for lunch) a horse-meat pasty. *Debaryomyces* has also been found on "an infected hand in Hungary," and, separately, the fingernail of a corpse.[31] *Cyniclomyces guttulatus* is more finicky and is difficult to find, growing, as it does, in the stomach lining of chinchillas.[32]

The ocean may seem an improbable home for yeasts when we think of their predilection for sugary fluids, and it is true that they do not flourish in the open sea where dissolved nutrients are scarce.[33] In estuaries and coastal waters, however, yeasts do very well, growing on seaweeds, and blooming in outpourings of nutrients from the mouths of rivers. They also voyage across the wine-dark sea in the guts of fish and marine mammals. Despite their numbers, marine yeasts are an underappreciated tribe, overshadowed by the photosynthetic plankton that energize the food webs that support lobsters, fish, walruses, and whales.

Their dismissal, even by oceanographers, is comparable to the disinterest with which most foresters view fungi. We see the fish and ignore the yeasts that fertilize the oceans, just as we admire the trees on land and overlook the mushroom colonies that sustain them.

In addition to the aforementioned *Debaryomyces,* marine yeasts with global distribution include another generalist called *Rhodotorula,* and *Candida* and *Cryptococcus,* whose landlubber relatives live on us and can cause serious diseases including brain infections. Marine yeasts grow on food particles suspended in the water and absorb dissolved nutrients. They use oxygen when it is available, and turn to fermentation when levels fall. As decomposers, their numbers are greatest in the polluted water around coastal cities, where they perform the essential task of breaking down the clouds of human waste that billow into the sea. *Metschnikowia* shows up in a lot of water samples too, often inside crustaceans and fish through which it cycles using its darts.

Yeast density decreases with distance from land, but they have been isolated from water sampled in the middle of the Pacific Ocean at a depth of four kilometers. Sequencing of DNA collected from the deep Atlantic, close to the wrecks of the *Titanic* and the German battleship *Bismarck,* shows that yeasts are more common than filamentous fungi.[34] This may have something to do with the dispersion of nutrients in the water and rarity of solid food particles amenable to hyphal colonization. Yeasts also show up in deep-sea muds and in forbidding environments like the Kuroshima Knoll, in Japanese waters, where methane and hydrogen sulfide gas bubble through cracks in the limestone seafloor.[35]

Bacteria and archaea—the simpler prokaryotes—are the champions of biology in extreme environments defined by the worst conceivable conditions that permit life. Microbiology is fantastic in its impudence, with a type of archaea from hydrothermal vents that grows in boiling water and stops doing so when "chilled" to 90°C, another that lives in a hot spring as corrosive as battery acid, and a bacterium that keeps four copies of its genomes on hand as an insurance policy against destruction by ionizing radiation.[36] The more complicated cells of eukaryotes, with multiple chromosomes boxed within nuclei, are never as hardy, but yeasts are as tough as eukaryotes get. Black yeasts, whose walls are impregnated with melanin, are often found in punishing environments. They live on rock surfaces and in crevices where they cope with intense ultraviolet radiation and precious little water, they can be found in brine pools saturated with salt, and they even festoon the walls inside the concrete sarcophagus that caps the ruined reactor at Chernobyl in Ukraine.

Yeasts are also some of the most heat-tolerant eukaryotes, with the aptly named *Candida thermophila*, isolated from Korean soil, capable of growing at 50°C (122°F).[37] At the other end of the temperature scale, yeasts are among the few eukaryotes that can cope with frigid water. An abundance of yeasts grow in the icy water that flows from melting glaciers.[38] Glacial water can carry lots of nutrients, which is surprising if one thinks of advertisements for bottled water showing crystal clear streams burbling from snow-capped mountains. In reality, meltwater is often clouded with "glacial milk" produced by the erosion of bedrock, and infused with organic matter derived from soil, plants, and microbes released from the melting ice. As glaciers thaw, this ancient organic matter pours into the outwash and feeds the yeasts. When the swarms of yeasts break this rich fuel down they are unlocking carbon that has been stored for thousands of years. Glaciers also contain microbes living in a state of multi-millennial suspended animation. Ice cores from Europe's southernmost glacier, the Ghiacciaio del Calderone, in the Apennines in Italy, contain hordes of yeasts primed for the end of their imprisonment.[39] With glaciologists forecasting the disappearance of this ice by 2020, these fungi will not have long to wait.

Permafrost is another resting place for ancient yeasts, although the sugar fungus is not one of them. Russian scientists have resurrected these microbes from ice cored from Siberian permafrost that formed three million years ago, during the Pliocene Epoch, when global cooling prevailed and ice sheets expanded over the poles.[40] *Rhodotorula* and *Cryptococcus* strains were cultured from the interior of these ice cores, accompanied by some filamentous fungi. Their genes aligned beautifully with the DNA of their contemporary relatives. Three million years represents a relatively brief sleep for these fungi, too short for their active descendants to have elaborated major evolutionary innovations.

Rocks exposed on mountains in the Himalaya and the Andes, and in Antarctica, provide another unlikely habitat for yeasts, but varied species of *Cryptococcus* and *Rhodotorula* are found wherever samples are taken.[41] Cold soils are brimming with them in the same areas. Man-made habitats attract yeasts too, with the familiar generalist species spotting the interior of washing machines in Slovenia, where a particularly thorough survey was conducted.[42] Indoor air is also filled with yeasts.[43] Some are carried in droplets of water, others are rafted around on dust particles, and the mirror yeasts shoot their spores into the air using their water droplets. Outdoor air is rich in yeast cells. Foggy conditions are very effective at moving them around, but they are always up in the sky, wet or dry, and circulate in the stratosphere, where they have been found floating at altitudes of 18–27 kilometers (11–17 miles).[44]

Mirror yeasts may be more important in the working of the planet than anyone imagined until recently. The dynamics of water movements in clouds are very complicated and there are a lot of unanswered questions about the early stages of raindrop formation via the condensation of water vapor. The technology of raincloud research is highly sophisticated, involving the use of laser-imaging instruments that are carried on aircraft flown into clouds. Condensation is promoted by the presence of solid particles in the air, and fungal spores provide a huge collective surface area for attracting water molecules. More than fifty million tons of fungal spores are dispersed into the atmosphere every year, corresponding to one million spores for every square meter of the earth's surface. The spores provide a total surface area of thirty-one million square kilometers (twelve million square miles), which is the same as the land area of Africa.[45]

Around half of this invisible mist comes from mushrooms and mirror yeasts that use the droplet mechanism to get airborne; the rest comes from filamentous fungi that spurt their spores skyward using hydrostatic pressure. Mirror yeasts can be isolated from outdoor and indoor air simply by removing the lid of a fresh culture dish, exposing its jelly on a window ledge for a few minutes, and putting the plate in an incubator. Mirror yeast colonies will always pop up in a day two, along with bacteria and filamentous fungi, suggesting that they are major occupants of the atmosphere.

Any particles can promote water condensation in clouds, but experiments using a special kind of microscope show that the spores of basidiomycetes have an undue influence on this process.[46] This instrument, called an environmental scanning electron microscope, allows exquisite control over the temperature and humidity of the air surrounding a sample of spores. When the spores are exposed to the high humidity levels that prevail in rainclouds, big water droplets grow on their surface like inflating balloons. The spores appear to be making their own raindrops. The explanation for this phenomenon is that the mechanism of droplet formation that powers the launch of the spores can be reactivated after the spores are dispersed. The details of this process suggest that fungal spores, including those produced by the mirror yeasts, may stimulate the birth of raindrops. *Bensingtonia, Sporobolomyces*, and company, may be rainmakers, giving them an oversized role in the health of forests and grasslands on which we depend.

Buller had no idea that his experiments on jumping spores were relevant to weather forecasting. Finding a connection like this is one of the joys of the life scientific, a buffer against the frustrations of lab work that can seem so futile at times. A great deal of research is demonstrably futile, of course, but it is difficult to predict this as a project kicks off. A trio of Nobelists would probably agree with this conclusion. Their work on the control of cell division in the 1970s and 1980s involved *Saccharomyces* and another yeast, *Schizosaccharomyces pombe*, also known as fission yeast (Figure 20).[47]

In the 1970s, Leland Hartwell discovered the cell division cycle (CDC) genes that are critical for cell division; Tim Hunt described a protein, called cyclin, which was produced in pulses whose timing preceded the division of cells, and Paul Nurse identified enzymes, called kinases, which seemed to be activated by cyclin. Hartwell worked with *Saccharomyces*, Hunt was studying sea urchin eggs, and Nurse adopted the fission yeast as his experimental organism. Mutant strains of fission yeast with defects in division brought it all together: *CDC* genes encode protein kinases; protein

Fig. 20. Scanning electron microscope image of the fission yeast, *Schizosaccharomyces pombe*.

kinases are activated by cyclins; and the activated enzymes guide cells through the division process. They were awarded the Nobel Prize in Physiology or Medicine in 2001.[48] Hartwell is also credited with the elucidation of cell cycle checkpoints that ensure that cells divide properly; Hunt went on to show that cyclins are produced by vertebrate animals; and Melanie Lee, who worked with Nurse, identified the cyclin-dependent kinases in humans.

With various bells and whistles, protein kinases activated by cyclins choreograph the cell cycle in all eukaryotes, from yeasts to humans. When this biochemical routine is disturbed, cells are given license to divide without their usual restraints. The repercussions can be catastrophic. Uncontrolled cell division leads to tumor growth in animals like us. Critics of the scientific enterprise who might be encouraged to belittle research on sea urchins and fungi, should show more caution in wielding their ignorance. Anyone who develops a tumor, whatever their political persuasion, may reconsider the benefit of investing in the study of these lowly creatures. The discovery of cyclin-dependent kinases has invigorated the search for drugs that control cell division and accompanying clinical trials to develop new cancer therapies.

Beyond their unicellular nature there is little resemblance between fission yeast and the sugar fungus. Apples and oranges provide an inadequate comparison. Apples and pine nuts offer a better match, because apples and pines belong to botanical groups that separated at least 300 million years ago, which may be close to the timing of the evolution of the classes of fungi that contain these yeasts.[49] There is a lot of uncertainty about the age of the gulf between the two yeasts, however, with some genetic comparisons suggesting that their common ancestor may have been a fungus that lived more than one billion years ago. This speaks to the fallibility of the molecular clocks used to date microorganisms, but we are left with the incontrovertible fact that *Schizosaccharomyces* and *Saccharomyces* are very different beasts.

The name fission yeast refers to the mechanism by which the cells divide. There are no buds, which means that we cannot identify a mother cell and her daughters. Fission yeast makes more cells by elongating and cutting its cylinders neatly in two through the middle. Schizo-comes from the Greek word *skhizo-*, which means splitting. The species is, therefore, the sugar fungus that splits. Pombe is Swahili for beer, and was applied to the fission yeast when it was isolated from millet beer in East Africa in the 1890s.

Schizosaccharomyces ferments sugars into alcohol using the same metabolic pathway as *Saccharomyces*, but it is no rival for the sugar fungus in brewing and wine making. Even in the natural fermentation of millet beer it is a minor player compared with *Saccharomyces*, which produces most of the alcohol.[50] There have been attempts to ferment grape mash using fission yeast to achieve a more controlled second fermentation, but the results suggest that *Saccharomyces* has very little to worry about. Indeed, the fission yeast is associated with spoiling wine by introducing "off-characters."

Schizosaccharomyces does have one metabolic party trick that has excited winemakers. This is the ability to turn malic acid into ethanol. In commercial winemaking, bacteria are used to reduce the acidity of the wine. This is not a perfect arrangement, because the bacteria can also introduce their own unpleasant flavors. Fission yeast has been substituted for the bacteria, with one company using a sort of giant yeast-filled tea bag as a dunk in the fermentation vessel. The beauty of this formulation is that the nylon bags and their fission yeast can be removed when the perfect level of acidity is reached.

Fission yeast has three big chromosomes that carry the same amount of DNA as the sixteen smaller chromosomes of *Saccharomyces*. Its complement of around 5,000 genes compares with the 6,000 genes of the sugar fungus.[51] Like the *Saccharomyces* genome, many genes in fission yeast have matches in the human genome. Fission yeast chromosomes have a similar structure to mammalian chromosomes, with large centromeres—often diagrammed like the hubs of propellers—and free ends, called telomeres, bulked up with proteins. *Saccharomyces* chromosomes have tiny centromeres and few proteins at their ends. Centromeres work as anchors for spindle fibers that separate the chromosomes when the cell divides. The process of cell division is another feature of the fission yeast that bears greater similarity to the animal mechanism than the uniquely fungal process of bud formation common to other yeasts.

Schizosaccharomyces is one of several microorganisms that ferment a popular tea drink called kombucha. The supposed health benefits of kombucha, which include improved digestion, mood elevation, and weight loss, have turned the bottled drink into a popular product, with annual sales in the United States reaching several hundred million dollars. Fission yeast, *Saccharomyces*, other yeasts, and accompanying bacteria occupy a slimy slab called a zoogleal mat that sits atop the fermenting tea. More disturbing than the buoyant remains of a large rodent, or an abandoned prosthesis, the presence of a yeasty kombucha medallion in a swimming pool would send bathers scrambling to their towels. The specific role of fission yeast in flavoring kombucha is unknown. Even if it were shown that it was responsible for the restorative powers of kombucha, one would be forced to concede that fission yeast's claim to fame lies in its lengthy career in biomedical research.

Most biologists have little knowledge of the yeasts; microbiologists who concentrate on bacteria and viruses offer scant coverage of these fungi in their textbooks, and even mycologists would rather look at mushrooms. Brewer's and baker's yeast are taken for granted, but the larger lives of yeast are easy to ignore. It takes scientific knowledge, as well as considerable imagination, to appreciate the skills that have enabled these fungi to sustain the planet, 24/7, for hundreds of millions of years. Imagination is an important part of the unfolding revelation—that tiny yeasts are big players in the workings of our dot in the cosmos. Even when we look at yeasts with a microscope it is difficult to equate the jostling cells with the well-being of the only place in the universe where we know that life exists. These fungi get on with things in silence, invisibly. Self-interest has attracted some inspection of the numerous microbes that grow on our skin, but the yeasts that claim greatest attention are those that affect our health directly.

7

~

Yeasts of Wrath

Health and Disease

Every gift from nature comes with a cost. Roses have thorns, kisses spread viruses, and the sugar fungus reveals the dark side of its personality when it enters our blood. *Saccharomyces*, mankind's benefactor and potential savior, belongs to the legion of opportunistic microorganisms that blossom in our bodies when our natural defenses are weakened. Infections, or mycoses, involving the sugar fungus are very rare, however. With fewer than a hundred reported cases in a decade, butterflies are a greater threat to public health.[1]

Most serious mycoses develop when our immune systems are impaired. People whose HIV infections are not responding to therapy and cancer patients whose defenses are weakened by radiation treatment are among the most vulnerable populations. *Saccharomyces* responds to these invitations and also finds its way past relatively intact immune systems through catheters implanted in patients admitted to intensive care units. The fungus also enters the body cavity during open-heart surgery. Opportunistic infections of these kinds are a modern phenomenon. The Roman gladiator impaled with a trident would not have survived long enough to be troubled by a yeast infection.[2] The technology of modern medicine encourages our colonization by fungi.

Once the yeast slides into the body via a catheter tube, or settles on a replacement heart valve, it multiplies if it can absorb enough food. The yeast spreads in the bloodstream, causing an illness similar to sepsis due to bacterial septicemia. "Fungemia" is the term for yeasts and other fungi in the blood. But blood offers slight rewards for a fungus that is fashioned to bathe in the sweet juice pressed from grapes. There is little evolutionary preparation on the part of the fungus for this struggle. We know this because the genes of yeast strains isolated from blood samples cannot be distinguished from those used in winemaking.[3] It seems that budding in the bloodstream is a matter of brute survival for a microbe that would be much happier in a wine barrel.

Diagnosis of fungemia can be difficult, because the presence of the yeast can be obscured by complications of underlying illnesses, but fever, sweating, nausea, and other flu-like symptoms are common. Because *Saccharomyces* has made no genetic investment in damaging the human body and grows in us so rarely, it shows little resistance to the standard armamentarium of antifungal medicines and is swept away without further ado in two-thirds of cases. Another of its failings as a disease agent is its inability to bury itself in solid animal tissues. In short, the sugar fungus is a poor opportunist.

Two of the published clinical reports of yeast infection say more about human desperation than the tendency of the fungus to cause disease.[4] Both concern instances of self-inflicted *Saccharomyces* infection. The first, from Columbia, Missouri, in the 1970s, involved a sixty-eight-year-old man with an unusual diet. He was a health food enthusiast who medicated himself with vitamins and, in a vein unrelated to his belief in dietary supplements, drank a full pint of vodka every day. Then he began developing symptoms of influenza and was admitted to a hospital. Nothing in the case history alarmed his physicians until the patient admitted to swallowing massive quantities of brewer's yeast. The clinical report says that he consumed up to three kilograms of dried yeast per day, which is equivalent to hundreds of the little sachets used in breadmaking. There is likely a typo in the transcript. He would have needed much more than a pint of vodka to wash that down. In any case, this gentleman developed fungemia that was misdiagnosed initially as a bacterial infection. His condition improved when he "was instructed to discontinue use of brewer's yeast."

The second case is more bizarre. It involved Vietnamese refugees in a Hong Kong detention center who injected themselves with yeast precisely in order to induce infections, and therefore to be admitted a hospital. One of the patients was a teenage boy admitted to hospital with convulsions. The other was a woman suffering from shock who presented with a serious abscess in her breast, presumably at the injection site. Once they were admitted to hospital they absconded. In the column of the table reserved for clinical outcome, the report says, "Seen running away." One hopes that these victims survived.

Concern about *Saccharomyces* infections has grown with reports that they are associated with the use of yeast as a probiotic, as was the case with the vodka drinker. Probiotics are microorganisms whose ingestion is thought to bestow health benefits. Probiotic yeast is sold in capsules as a treatment for a wide range of digestive disorders, and as a daily supplement to maintain a healthy bowel function. The capsules contain a freeze-dried preparation of *Saccharomyces boulardii*, which is probably a strain of *Saccharomyces cerevisiae*, rather than a separate species.[5] Nonetheless, it behaves quite differently from the strains of the sugar fungus used to brew beer and leaven bread. The association between the use of *boulardii* as a probiotic and development of yeast infections should not concern the majority of people who have benefited from the probiotic.

Probiotic yeast was discovered in 1923 by French microbiologist Henri Boulard, who went upriver somewhere in the colonial territories of French Indochina. Rather than searching for Kurtz,[6] Boulard's quest was for a yeast that could ferment wine in warm climates. The story gets a bit murky, but it seems he isolated the yeast from the skin of lychees (some sources refer to mangosteens too), which locals chewed as a remedy for diarrhea caused by cholera. Boulard may have developed cholera himself during his travels, and resorted to drinking tea brewed with the yeast while he writhed on a filthy cot inside a mosquito net.

Today we find *boulardii* capsules advertised in glossy magazines as a palliative for irritable bowels. Henri Boulard sold the patent for his yeast to a French company called Biocodex, which began manufacturing *boulardii* for use as a probiotic in the 1950s. Biocodex has maintained the patent on the original yeast strain and markets the product using the trade name Floraster®. Other formulations of *boulardii* are sold by the global leaders in yeast manufacture for the brewing and baking industries, including Lesaffre and Lallemand.

The global probiotic business has been corrupted by poor product control and unsubstantiated claims made by advertisers.[7] In the absence of the governmental regulation that applies to prescription medications, probiotic capsules and drinks like kombucha are advertised as remedies for everything from tummy upsets to colon cancer. But in this cesspool of pseudoscience, *boulardii* floats to the surface because there is compelling evidence for its effectiveness at treating specific non-life-threatening conditions.[8]

Intestinal distress is a common condition in patients prescribed antibiotics that wipe out a broad range of bacteria. In some cases, bacteria that can resist the antibiotic therapy grow at a fantastic rate, overwhelming the remaining microorganisms in the gut, and destabilizing the whole digestive system. *Clostridium difficile* is the commonest of these pests. It releases toxins that cause the bloating and severe diarrhea described as colitis. *Boulardii* controls these symptoms by modulating the response of the immune system within the gut.

Immune-boosting properties are claimed for every probiotic on the market, but, uniquely, they are backed by the results of controlled experiments for *boulardii*. Treatment with probiotic yeast also helps encourage a healthy mix of microorganisms to recolonize the gut after a bout of colitis.[9] The yeast may also be useful for treating diarrhea in infants, and to relieve the commonplace suffering of travelers whose sensitive digestive systems are made playful by water contaminated with bacteria. Beyond these ailments, evidence for the effectiveness of *boulardii* is very limited, but this does not hamper the online marketing of the capsules to those afflicted by irritable bowel syndrome, Crohn's disease, cystic fibrosis, and many other illnesses.

The concept of probiotics was the brainchild of Élie Metchnikoff of cellular immunity fame.[10] As a young man in Russia he had become interested in a drink called "kumis," or "koumiss," made by fermenting mare's milk. Taking the "kumis cure" had become very fashionable at the end of the nineteenth century, and Russian resorts dedicated to the therapy attracted people with bronchitis, tuberculosis, and other illnesses. Tolstoy became an early devotee, confident that kumis alleviated his depression, and Chekhov embraced the cure for his tuberculosis, from which he died at the age of forty-four. After his groundbreaking work on phagocytosis by predacious yeasts in the water flea, *Daphnia*, Metchnikoff moved to the Pasteur Institute in Paris, at the invitation of its namesake. In France he started drinking sour milk every day in the belief that it stimulated digestion, and a colleague at Pasteur introduced him to Bulgarian *yahourth*, or yogurt. This coagulated dairy product was a staple food among peasants and Metchnikoff became convinced that there was a causal connection with their purported longevity. He thought he had discovered the fountain of youth.

In 1904, Metchnikoff delivered a public lecture, "La Vieillesse" (Old Age), in which he urged the audience to avoid raw food, because it was contaminated with germs, and to consume yogurt to reduce the effects of harmful bacteria in the colon. This message grabbed the attention of newspapers and launched a craze for sour milk products, which were used to treat diarrhea in babies, digestive problems in adults, and the chronic illnesses of one's dotage. Metchnikoff believed that the human lifespan could be extended to 150 years, and when he won the 1908 Nobel Prize for his immunological work the demand for sour milk soared. He died from heart failure the next year at the age of seventy-one.

A century later, yogurt is a multibillion-dollar darling of the food industry, and many of the most popular brands are marketed as health foods. A bestselling brand called Activia®, owned by Groupe Danone, had been marketed as a proven relief for irregular bowel habits until the company was forced to mute its claims by the Federal Trade Commission and pay millions of dollars to settle a class-action lawsuit by dissatisfied customers.[11] Activia and other yogurts are bacterial products in which yeast growth is discouraged by the acidification of the milk during the fermentation process. There is some interest in adding probiotic *boulardii* after the fermentation is completed to augment the health claims made by manufacturers. Unlike its baking and brewing counterparts, *boulardii* is quite acid-tolerant, which makes it possible that before long we will be treated with a new generation of advertisements for anti-diarrheal yogurt. Kefir is another fermented dairy product attracting interest from companies investing in probiotics.

The human colon is dominated, when we eat well, with the fibrous remains of vegetables, and is no place for *Saccharomyces* with its cravings for an immediate sugar fix. The sugar fungus is not a major player in the conglomeration of live microbes that make a living in our inner tubing. Its genetic signature is often present in human feces, but it rarely grows in cultures made from scrapings from the gut mucosa taken during colonoscopies or from stool samples.[12] This suggests that it is moving through the human intestine in the form of dead cells ingested in bread, beer, and other fermented foods. Yeast has been shoved through the gut in this funereal procession since we adopted it as our dietary helpmate thousands of years ago. This has happened, of course, in less than an eye-blink in evolutionary history, and the bacteria that have lived inside us from the beginning of our species have had to develop ways to work with this newfangled fungal debris in short order.

One of the most abundant intestinal bacteria, *Bacteroides thetaiotaomicron*, specializes in the breakdown of the branching chains of a sugar polymer called mannan that sticks out from the surface of the yeast cell wall.[13] It performs this considerable metabolic feat—and nothing else seems to have learned how—in a self-interested fashion that leaves no scraps for other microbes. The intact mannan molecules are shaped like trees. If the bacterium snipped these chemical structures in a haphazard manner to get at the individual sugars, juicy bits of the molecule would be released to other microbes in the surrounding fluid. *Bacteroides* pares away at the structure with such surgical precision that nothing is left for other bacteria. Selfishness pays by allowing this tidy eater to harvest all of the yeast polymers for itself.

In a handful of medical oddities, living cells of the sugar fungus do manage to grow in our digestive system, which is a bad thing because they turn their hapless vessels into the most intimate of home breweries. This phenomenon, known as "gut fermentation syndrome," or "auto-brewery syndrome," is even rarer than cases of blood infection by the sugar fungus. A case in Texas involved a man who experienced severe intoxication on many occasions after drinking little or no alcohol.[14] Without the stimulus of an all-night bender, his blood alcohol level often rose between 0.3 and 0.4 percent. This is sufficient to cause stupor, and slightly higher levels are considered life-threatening. The legal limit for drivers is 0.08 percent in the United States, as well as in England and Wales, unless one is operating fixed-wing aircraft, in which case the level is reduced to 0.02 percent. Auto-brewery syndrome was confirmed in the Texas patient. *Saccharomyces* was growing in his gut, fermenting sugars in his diet to produce alcohol and CO_2. This poor man got drunk without drinking. Treatment was straightforward. Following a six-week course of antifungal drugs, the yeast disappeared from his intestines and he resumed a life of sobriety.

Similar symptoms have been reported in children with short bowel syndrome, which results from the surgical removal of a damaged part of the intestine or occurs as a congenital condition. The appearance of auto-brewery syndrome in these cases supports the idea that yeast growth is promoted by abnormalities in the digestive system rather than vice versa. The origin of comparable changes in gut function in adults is a mystery but the condition has been introduced as a defense against charges of drunk driving. This is a poor alibi on most occasions because the levels of alcohol produced by yeast are so low, but it resulted in the acquittal of a woman in New York in 2015 whose blood alcohol level was four times the legal limit.[15] The syndrome has some interesting legal and theological implications for countries and religions with zero tolerance toward alcohol.[16]

Experiments have demonstrated that low levels of alcohol are produced in most of us when we consume sugars.[17] According to this research, intoxication by intestinal yeast is an extreme manifestation of a universal feature of human digestion. Because fermentation is only possible for live yeasts, this conclusion is at odds with the absence of *Saccharomyces* in cultures obtained from stool samples. The explanation may be that other yeast species turn glucose into ethanol. Uncertainty reigns because the gut microbiome, particularly the fungal part of it, remains an uncharted wilderness for scientists. Irrespective of the source of these traces of ethanol, the phenomenon of onboard fermentation raises some tantalizing possibilities about constitutive alcohol exposure in children and our varying appetites for cocktails in later life.

Alcohol tolerance in adulthood may also have something to do with variations in the microbial inhabitants of the gut, and daily pulses of alcohol from the microbiome could affect our level of alertness and even make us a little happier. We tend to ascribe daily rhythms of happiness to hormones, neurotransmitters, and positive thinking, but fermentation by intestinal yeasts may be important too. The schoolchild with the sunny outlook and infectious laugh may be getting a kickback from her microbiome. A designer yeast with such euphoric properties would be a very attractive probiotic indeed.

While this hypothetical dribble of alcohol from yeasts in our digestive system might be a good thing, the presence of *Saccharomyces*, as whole cells, dead or alive, and as fragments, is also associated with the development of inflammatory bowel diseases (IBDs). We know this because the formation of antibodies against yeast is useful in the diagnosis of Crohn's disease. Crohn's is one of the major forms of IBD, whose common symptoms include abdominal pain, diarrhea, fever, and weight loss. The anti-*Saccharomyces cerevisiae* antibodies detected in the blood react with the mannan polymer on the yeast surface. More than half of patients have these antibodies and yet they are rare in people without the illness.[18] This means that the blood test misses a lot of sufferers, but also has a low rate of false positives. Surprisingly, healthy family members of Crohn's patients are also more likely to produce antibodies to yeast mannan. This curious phenomenon is probably explained by a genetic predisposition to the disease, because spouses of Crohn's patients do not test positive for yeast sensitivity.[19] Like other autoimmune diseases, susceptibility to Crohn's may be a by-product of an advantageous response to parasitic infections way back in our evolutionary history.

The link between the antibody and the illness is unknown. Because the wall of the intestine becomes inflamed in Crohn's disease, it is thought that the resulting thinning of the protective mucus lining allows bacteria that are normally swept downstream to pile up on the damaged tissue. Yeast mannan in the gut that leaks through these sore spots may be responsible for inducing antibody formation. Mere presence of the antibody does not mean that yeast plays any active role in producing the symptoms of this form of IBD. It is quite possible that the antibodies are akin to bystanders surrounding a car crash: like the spectators, yeast is not responsible for the damage.

Celiac disease comes with many of the same symptoms as Crohn's disease, but it is a distinctive illness stimulated by immune reactions to gluten in cereal grains. Gliadin, which comes in three forms, is the problematic component of gluten and its elimination by adopting a wheat-free diet is the simplest remedy for celiac symptoms. Celiac is an autoimmune disorder in which the finger-like villi that do the job of food absorption in the small intestine are attacked by the patient's immune system. Fewer than 1 percent of the human population has this condition, but recognition of the effectiveness of ridding gluten from the diet of celiac patients has spawned a craze for gluten-free bread in the United States.[20] Bearers of the brumous job title of "nutritionist" have suggested that modern cultivars of wheat produce grain with high levels of gluten, and that this has become a poison within the diet.[21] Scientific evidence for this conclusion is lacking.

Some studies have indicated that celiac disease, like Crohn's, is associated with the formation of antibodies against yeast and that the antibodies disappear after patients exclude gluten from their diets.[22] The same anti-yeast antibodies have also been reported in patients suffering from rheumatoid arthritis, type I diabetes, systemic lupus erythematosus, and other autoimmune disorders.[23] These findings may point to a larger correlation between inflammation of the gut and an immune response against *Saccharomyces*, but there is no proven cause and effect relationship between the sugar fungus and any of these serious illnesses. Nevertheless, a caucus of nutritionists has coopted these uncertainties and characterized yeasts as the enemy of the inflamed bowel.

In modern America—where sales of anti-diarrhea pills, laxatives, and stool softeners suggest that an entire society is obsessed with its collective intestinal peristalses, or lack thereof, and where it is undeniable that we are being rendered obese and diabetic from eating too much terrible food, and becoming increasingly resistant to scientific facts and, more generally, unhappy with life—a mélange of half-truths and complete lies has spawned an industry of dietary misinformation that encourages people with real illnesses like Crohn's and celiac disease, as well as people sickened by nothing more than their own gullibility, to give up wine, beer, bread, soy sauce, and every other yeasty food. The Internet is rife with sermons on this topic, and books including *The Yeast Connection: A Medical Breakthrough*, written by the late, and aptly named, Dr. William G. Crook, are persistent bestsellers.[24]

Contrary to fashionable efforts to rid all yeasts from the diet, probiotic *boulardii* yeast is still recommended by nutritionists to treat the diarrhea associated with IBDs. This does not make sense, given that the sugary mannans considered so dangerous by prophets of yeasty doom are just as numerous on the cell surface of *boulardii* as they are on other strains of yeast. According to these practitioners of alternative medicine, there are good yeasts and bad yeasts, except when all of them are bad—or not. One can be forgiven for feeling confused. People enjoying reasonable healthy lives could do worse than calling upon the gastroenterological wisdom of my Lincolnshire grandmother, a woman who attended to intestinal health with alarming enthusiasm. Her morning greeting was, "How are your *doings* dear?" Unless one reported excellence in this department, she would prognosticate with a sureness that spoke to decades of medical training, to which she had no formal claim. She would not have had any truck with probiotic remedies when the quotidian ills of the abdomen could be addressed with a sensible diet and some fresh air.

Fresh air in Lincolnshire, however, and everywhere else, is often brimming with fungal spores. Many of these particles come from yeasts, and they cause asthma. Because the sugar fungus is an earth-bound microbe that cannot move anywhere on its own, it does not make it into our nasal passages and lungs very often. Bakers are more likely to inhale *Saccharomyces* cells than the rest of us, but the proteins in cereal flour pose a greater occupational hazard. Baker's asthma caused by bakery dust was recognized by the Romans and described by the Italian physician Bernardino Ramazzini in the 1700s.[25] Yeast has been reported as a cause of the illness when it is used in its dry form rather than the conventional cream yeast used in commercial bakeries, but home bakers and brewers need not be worried at all.

Other yeasts, particularly the mirror yeasts like *Sporobolomyces* that use water drops to get airborne, are a big problem for asthmatics—more important, perhaps, than plant pollen. A medical detective story reported in *The Lancet* in the 1980s began with the admission of seventy patients to hospitals in Birmingham, England, following a summer thunderstorm.[26] Measurements showed that levels of pollutants from industrial plants and cars were comparatively low around the time of the cluster of asthma cases, and fluctuations in pollen levels did not correspond to the event either. The smoking gun was discovered in detailed analysis of the concentrations of fungal spores during the thunderstorm. As the band of rainclouds moved over the West Midlands of England, *Sporobolomyces* budded and launched spores with unusual ferocity, blanketing the city with 500,000 spores per cubic meter of air. The National Allergy Bureau in the United States categorizes counts above 50,000 spores per cubic meter in the "very high" category, emphasizes the numbers in red in their daily reports, and suggests that people with asthma stay indoors.[27]

The mirror yeast stimulated by the summer storm in England was rebounding from a period of hot dry weather that had curtailed its growth and prevented spore discharge. The downpour rejuvenated the fungus, allowing it to multiply by budding, and the high humidity levels stimulated the drop formation that flicked its spores into the air. Rainfall scrubs spores from the atmosphere, which makes the immense numbers of airborne *Sporobolomyces* cells detected in the Birmingham study so impressive. Breathing difficulties associated with stormy weather cause considerable suffering among the hundreds of millions of people with sensitive airways and is called summer asthma. Many kinds of fungi and other microorganisms are stirred by rainfall, but the ubiquity of the mirror yeasts, coupled with their astonishing numbers, suggest that they are among the most important environmental causes of asthma. They are the number one enemy of the wheezy child. The addition of the proven role of airborne yeasts in asthma to their putative abilities in rainmaking raises the profile of these fungi as grand players in the biosphere.

Asthma is a widespread and potentially life-threatening condition. Estimates suggest that 300 million people suffer from the affliction and 250,000 deaths are attributed to the illness every year. Allergy testing proves that fungal spores are among the commonest causes of asthma. Starfish-shaped dendritic cells police the lungs, conveying fragments of proteins from spores and other foreign materials to their surfaces and presenting them to other kinds of cells in the immune system. This process oversensitizes the lungs of asthmatics, so that repeated exposure to particular irritants causes the release of inflammatory compounds and overproduction of mucus that result in airway constriction. Inflammation of the nasal passages is also caused by the inhalation of spores. This condition is called "allergic rhinitis," or hay fever when it occurs in spring or summer. Allergic alveolitis is a third illness that can be caused by exposure to high concentrations of airborne yeast cells. Alveolitis affects the deepest lung tissues where air swirls into the microscopic air sacs or alveoli. *Sporobolomyces* cells are much smaller than the spores of many fungi—perfect for passing all the way down into the alveoli. A case of alveolitis caused by this yeast was reported in an equestrian exposed to contaminated bedding straw in a horse barn.[28]

Like baker's yeast, *Sporobolomyces* and other mirror yeasts are opportunists whose occasional growth in human tissues is a source of great interest for experts in medical mycology. It must be difficult for some of them to hide their excitement when they are confronted with a patient transformed into a living Petri dish, feeling some empathy for the feverish man in the bed, perhaps, but thinking already about the way that his fungus is going to enliven their PowerPoint presentation at the next annual meeting. Published case histories include nasal polyps and skin blisters filled with the yeast, dermatitis, and deeper infections beneath the skin, in the lymph nodes, and in multiple sites around the body of patients whose immune systems have collapsed.[29] Cancer therapy, infection with HIV, and intravenous drug use are the common invitations to the fungus. The yeast has also been identified in otherwise healthy patients, including a young woman with *Sporobolomyces* growing in the vitreous of her left eye.[30] Cases like this defy simple explanation. How had an airborne yeast made its way into the jelly inside this woman's eye? In the majority of cases, yeast infections respond to treatment with the standard antifungal medicines, but without a functioning immune system fungi will find their way back into the unprotected tissues.

Sporobolomyces has no particular appetite for human tissues, and its broad host range includes salmon fry in fish farms, as well as dogs and other terrestrial pets. To this microscopic yeast, indeed to most of nature, there is nothing to choose between the dog walker and her pooch. Both menu items are organized as mineral frames strung with bands of protein and smoothed with blobs of fat, electrically wired, aerated by bellows in the chest, nourished and drained via an elaborate plumbing system, appended with organ meats, and wrapped in an elasticated hide. When the internal organs of the woman or her basset hound are exposed, or the defense system of either mammal is on its knees, the great mass of microbes will attempt to grow in the warm mucus. Most cannot find what they need in this glistening fluid and give up the fight; a few, like the mirror yeasts, can scratch a meager living for a while. They would rather be somewhere with a richer supply of sugars and properly aerated. Life in the middle of a hardening lymph node is far from ideal for a fungus accustomed to growing on surfaces and free to toss its spores into the air. But within the encyclopedia of yeasts, a handful of species stand out for their readiness for the vicissitudes of life inside the human body. These fungi possess a suite of adaptations that suit them for combat.

This brings me to *Candida albicans*, the fungus known as vaginal yeast, which is a limiting nickname for a microorganism that also lives in our mouths, survives the acidity of the stomach, and colonizes the 6.5 meters (twenty-one feet) of intestine, where it grows as the most abundant fungus of the gut microbiome (Figure 21). The fictional role of this fungus in a catalog of health issues, as

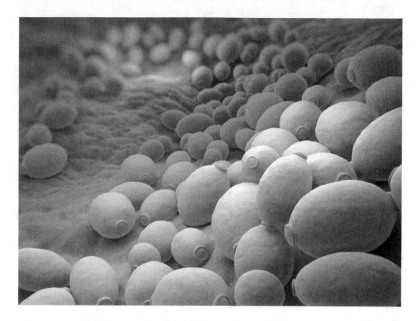

Fig. 21. *Candida albicans*, the vaginal yeast.

well as its actual manifestation as a distressing vaginal irritant, make *Candida* the best-known yeast after the sugar fungus. It is a much closer relation to *Saccharomyces* than either microbe is to the mirror yeasts or to fission yeast, and they are classified in the same family of the fungi. Despite this affinity, baker's and vaginal yeast belong to ancestral lines that separated more than 200 million years ago, budding at a time when herds of ichthyosaurs rushed beneath the whitecaps of the ancient seas.[31] Long after the extinction of its Triassic ancestors, *Candida albicans* acquired its tricks for infecting humans.

Yeast experts recognize more than 300 species of *Candida*, starting with *Candida aaseri*, isolated from sputum coughed up by a Norwegian, and ending with *Candida zeylanoides*, also discovered in Norwegian sputum. This shared site of domestic microbiological bliss speaks to a detailed study of patients with respiratory illness rather than a surfeit of phlegm in Scandinavia or a fungal attraction to Norwegians. Both species grow in other places and on other things. Indeed, these and other *Candida* species have been isolated from soil, seawater, pickled cucumbers, wild mushrooms, "turbid formalin in tanks in an anatomy department," a corpse in Switzerland, bees, and the gut of a blind click beetle (which is very different from a click beetle that has lost its sight).[32] Besides a handful of strains isolated from flowers and fruits, *Candida albicans* limits itself to the human body. It is on intimate terms with every one of us.

It evolved, most likely, as a benign member of the human microbiome, growing throughout the digestive system and in the vagina, feeding on mucus and growing in harmony with bacteria. From the platform of this supportive symbiosis, it also spread in a damaging fashion when its human hosts aged and sickened and the healthy mixtures of bacteria became unbalanced. We provoke the same behavior today when we take a course of antibiotics that wipes out the bacteria and releases the natural brake on *Candida* growth. The resulting proliferation of yeast causes vaginal inflammation and is an example of dysbiosis. An example of dysbiosis on a macroscopic scale is the population explosion of white-tailed deer in North America in habitats where large predators have been eradicated by hunting.

More serious conditions arise when *Candida* becomes an opportunist and penetrates the wall of the intestine or vagina. This happens when the immune system is compromised. The vulnerability of the body when the shield of the immune system is lowered illustrates the precarious nature of existence. Every second of our lives, our fate hangs on the cooperation between the trillions of cells that fashion our tissues and the legions of foreign organisms populating the microbiome. When our defenses are lowered, *Candida* sticks to the layers of cells beneath the protective mucus, stops growing as a yeast, and transforms itself into a filamentous fungus that buries itself in our organ meat.[33] *Candida* is dimorphic, or, if we quibble about intermediate stages between yeast and filament, may be described as polymorphic or polyphenic. The formation of filamentous hyphae changes everything for the fungus in a mechanical sense because it can begin thrusting ahead with the pressurized tips of its cells rather than pushing against its surroundings in all directions as a budding yeast. As this invasive growth proceeds, feeding occurs via the release of digestive enzymes from the filaments.

The result of this change in form can be catastrophic for the patient, with major damage to blood vessels, heart muscle, eyes, and central nervous system. In its filamentous phase it can even lay waste to the skeleton. These are the worst manifestations of "invasive candidiasis," a disease that affects 250,000 people every year and causes more than 50,000 deaths.[34] Candidemia refers to the infection of the bloodstream that occurs in patients after surgery or is associated with the use of catheters. Yeast cells spread in the bloodstream are responsible for the development of the deep infections that carry the highest mortality rate. *Candida* has developed resistance to some of the drugs used to treat fungal infections, but broad-spectrum antifungal agents called echinocandins remain effective at clearing the fungus if the treatment begins in the early stages of disease.

Treatment of candidiasis is complicated by the growing incidence of infections caused by other *Candida* species with different levels of resistance to the drugs. This places a premium upon the swift identification of the particular fungus responsible for each case of the disease. *Candida auris*, which was first isolated from a patient with an ear infection in Japan in 2010, is spreading at an alarming rate and has the worrying characteristic of resisting all of the available antifungal agents. Rather than developing an infection from microbes that spread from their normal occupancy of the gut microbiome, *Candida auris* seems to move from patient to patient in hospitals.[35]

The majority of cases of invasive candidiasis occur in older patients with serious underlying illnesses, or in people with weakened immune systems. *Candida* can subvert the phagocytes of the immune system—the cells discovered by Metchnikoff—by surviving after they are swallowed. In a normal immune response, microorganisms are destroyed inside the phagocytes, but *Candida* manipulates the immune cell, transforming its prison into a sanctuary. After a while, the yeast turns filamentous, pushes its way out, and kills the phagocyte, reversing the roles of predator and prey.[36] Even when the fungus deploys this devilish behavior, healthy immune systems are effective at detecting and destroying yeast cells that turn nasty. This keeps the yeast under control in our mouths and innards. And most of us have little to fear from *Candida*, unless we are admitted to hospital for a surgical procedure, require a deep catheter, or are robbed of immunological power by a virus, cancer therapy, or plain old age.

This reasonably good news is swept aside by Ann Boroch, bestselling author of *The Candida Cure*, which offers a ninety-day nutritional program to "restore vibrant health."[37] A table in her book lists more than eighty "[c]onditions caused directly or indirectly by [yeast] overgrowth," which include ALS (Lou Gehrig's Disease), HIV/AIDS, leukemia, mitral valve prolapse, obesity, alcoholism, suicidal tendencies, and sexually transmitted diseases. Like the late Dr. Crook, Boroch contends that deliverance from these illnesses is available by adopting a yeast-free diet.

The vilification of *Candida* by health food gurus has arisen from the misinterpretation of the fact that this yeast can cause serious infections. It is true that the yeast can become very troublesome, but it is a perversion of all logic, common sense, and science to link it with life-threatening diseases other than invasive candidiasis. In only one instance has science illuminated a potential link between a *Candida* infection and a health issue other than a *Candida* infection. Celiac disease, as we know, is associated with immunological sensitivity to the protein in wheat gluten called gliadin. Investigators have found that a protein generated by *Candida* filaments during tissue invasion is so similar in its molecular structure to gliadin that the immune system may have difficulty in telling the two proteins apart.[38] This raises the possibility that an immunological reaction to a *Candida* infection in the gut might lead to the development of celiac disease. This association is highly speculative, but certainly raises the possibility that the condition is founded on something other than an oversensitivity to wheat gluten, which has, after all, been part of our diet for thousands of years. This contribution to celiac research does not mean, of course, that there is any scientific basis for the all-encompassing claims against yeast by the alternative medicine community.

Fungal biologists studying candidiasis have been searching for virulence factors that account for the damaging effects of the fungus for many decades. "Virulence factors" are features of the yeast cells that help them stick to tissue surfaces, convert themselves into those invasive filaments, evade the immune system, spread in the bloodstream, and destroy solid tissues. Genes that facilitate these processes have been identified, but, increasingly, we recognize that the pathogenic behavior of the fungus is a consequence of immunological damage. The primary cause of candidiasis is immunological, which means that new treatments for these infections are most likely to come from therapies that support the immune system. Antifungal drugs can work wonders, but the relief will be temporary unless the patient regains some natural defenses. Drugs that target particular molecules within the fungal cell have been very effective at treating candidiasis, but with an increasing number of invasive infections and the phenomenon of multidrug resistance, immunotherapy seems to offer the best hope.[39]

We might consider the disease process from the viewpoint of the fungus. Submerged in the wall of the gut or a blood vessel, *Candida* has little chance of long-term survival. Like the sugar fungus, *Candida* does not produce airborne spores, which limits its options for dispersal. Once the patient dies, it has no way to escape the body and it is likely to decompose with the corpse. In nature, there is a possibility that the fungus will find its way into the gut of a scavenger, and the ability to cycle through the bodies of different animals may be an effective survival mechanism for some fungi. But disease is not the aim in life for *Candida*. It works best as a peaceful resident of the microbiome, shepherding bacteria and maintaining the healthy chemistry of the gut and vagina.

Our relationship with this yeast begins in infancy, when we are colonized by the strains picked up from the birth canal. Babies often develop a white overgrowth of the yeast in the mouth and covering the tongue, known as thrush. This common manifestation usually disappears in a few days. *Candida* continues to live in our mouths with less fanfare, and we probably swap strains with one another throughout our lives. It is in us now, feeding and budding, living its yeasty life.[40] Problematic growth of the fungus in candidiasis is part of what we might call the noise of nature, created by the occasional and inevitable imbalance between host and microbe during our decades of cohabitation.

Yeasts capable of the dimorphic switch from bud to filament are quite common and many of them use this as a survival mechanism when they become trapped in the bodies of animals. Histoplasmosis and blastomycosis are illnesses produced by environmental yeasts that undergo this makeover in the human body.[41] The fungi that cause these illnesses, called *Histoplasma* and *Blastomyces*, enter the body as airborne spores and spread from the lungs to the vital organs. They can even grow in bone tissue. Little is known about their habits in the outside world, but *Histoplasma* grows in soil enriched with guano from birds or bats. Both infections are concentrated in North America. Histoplasmosis is also known as Ohio Valley disease for its prevalence in my part of the United States. Blastomycosis is more widespread. Histoplasmosis is associated with exposure to guano in chicken houses or masses of bird droppings deposited by starlings or other wild birds that flock in large numbers. The majority of cases of histoplasmosis and blastomycosis go unnoticed; they occur as subclinical infections. Serious infections follow the usual pattern of taking hold of patients with impaired immune systems and are associated with high rates of mortality. Their rarity is a blessing.

Black yeasts whose cell walls are impregnated with melanin also show up as infrequent pests, and can do so, unfortunately, in patients with healthy immune systems. *Exophiala dermatitidis* is a dark-skinned thermophile—a heat lover—that lounges in saunas, steam baths, and dishwashers.[42] It presents itself as a skin infection, but has the unpleasant characteristic of spreading to the central nervous system where it causes brain infections. Five women in North Carolina developed this illness in 2002 following spinal injections with a steroidal medicine for pain management that was contaminated with the fungus. Four were cured with antifungal drugs, but one of the patients died from meningitis. Species of red-pigmented yeasts are also associated with serious infections. *Rhodotorula* is very widespread, discoloring shower curtains and grout around tiles in showers along with pigmented bacteria. It pursues a cheerless agenda as a pathogen, attacking immunocompromised patients and growing around implanted catheters.

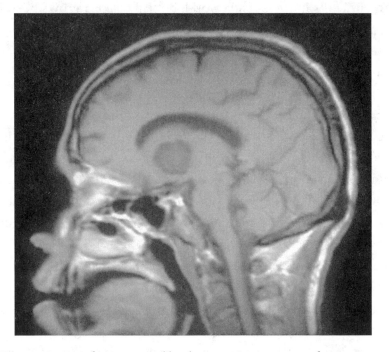

Fig. 22. Brain infection caused by the yeast *Cryptococcus neoformans*, in a patient with AIDS.

Finally, the deadliest of all yeasts belong to the genus *Cryptococcus*. These species produce filamentous colonies in soil and release tiny airborne spores. The spores are inhaled into the lungs where the fungus switches to a budding form that spreads in the bloodstream and specializes in brain infections (Figure 22). Cryptococcosis, caused by *Cryptococcus neoformans*, is a particular problem for AIDS patients in which it distinguishes itself as the most common life-threatening fungal infection. In advanced cases of the illness, patients suffer nausea, headache, light sensitivity, blurred vision, and confusion. Death follows these symptoms of fungal meningitis. A closely related species, *Cryptococcus gattii*, rounds out the horror by attacking people with perfectly serviceable immune systems. Most cases of the *gattii* infections occur in Australia and New Guinea, but lethal outbreaks have also occurred in North America. Because antifungal drugs are of limited usefulness, surgeons have been forced to excise the part of the brain carrying the fungus to save some of their patients. *Cryptococcus* is a loathsome microbe—a different beast, in every way, from the sugar fungus, our redeemer.

These examples of the cheerless manifestations of the fungi afford a stark contrast with the benign nature of most yeast species. This is the way of so many human interactions with microbiology. Everything works well until our immune defenses are lowered, or we encounter a rare pernicious germ in a way that the customary rapport is breached. Peaceful coexistence is the norm. A plenitude of gentle microbes live on our skin and eek out their existence without threatening ours. Yeasts are the commonest fungi on the two-square-meter surface of the average adult. The budding lifestyle seems to suit this challenging environment where expanses of dry desert are interspersed with oases of sweat and greasy sebum. Ear canals and nostrils are good lodgings and the scalp is a veritable pleasure garden for yeasts. Multiple species of the yeast called *Malassezia* are the winners in all of these locations, with up to ten million individual cells on every head—quadrillions of the little spheres lofted around the planet as *Homo sapiens* goes about its business.[43] *Malassezia* is a peaceful microbe in these locations, part of a healthy microbiome. It is connected with dandruff, growing with greater enthusiasm on the scalps of people with this condition and boosting sales of shampoos that stop it doing so. In darker moods, our companion causes serious forms of dermatitis and eczema, inflames hair follicles, and invades hair shafts.[44]

Malassezia is also found off-piste in deep marine sediments, hydrothermal vents, corals, lobster and eel guts, Antarctic soils, nematodes, and orchid roots.[45] We infer the abundance of the yeast in these locations from its genetic footprints, but we know very little about its lifestyle. Research on *Malassezia* is difficult because it resists cultivation in the lab, but the range of habitats populated by the fungus speaks to a group of microbes of immense ecological consequence. It does not seem surprising that a yeast that grows around hydrothermal vents may hold some scientific secrets, but the clandestine occupations of microbes on our skin speaks to a remarkable apathy on the part of researchers until very recently.

We know so little about our ecology, about ourselves. Each time you rub the smooth glabella between your eyebrows, yeasts are ferried on your fingertips. Getting up from your desk to shake hands with a visitor, the yeasts become her property, and hers become yours. Idle head scratching moves things around and new mixtures of cells are smeared on the buttons of your computer mouse. Every movement, each physical transaction, is traced invisibly in yeast DNA. Primates with sufficient knowledge and technology to sling spacecraft into orbit around distant planets do so from a world of mycological magic and mystery.

Yeasts come in many species, but the sugar fungus reigns supreme as our partner in civilization. We would not be here without her. *Saccharomyces cerevisiae* and *Homo sapiens* have been inseparable for thousands of years. We are reflections of each other, our genetic similarities reflecting the deep ancestral root from which our common cellular machinery arose. Matched expressions of these genomes allow the fungus to ferment alcohol and us to digest it. This metabolic coordination, spread over a few thousand generations of human pleasure and pain—alcohol delivering both—developed in the rainforests from which apes with an upright gait migrated to the savannah. Our complex relationship with alcohol, and later with leavened bread, drove agriculture and settlement. From these splendors came civilization, political organization, militarization, and mass starvation. Later fruits of our yeast-driven civilization included science and technology, engineering and medicine, exponential population growth, and the attendant destruction of the biosphere. And in this time of considerable climatic peril, industrial applications of yeast promise major advances in biotechnology and offer some hope—perhaps our only hope—of powering a carbon-neutral economy. The future of humanity depends more on this bug than on any farm animal or crop plant.

In short order, science has transformed the mysterious agent of fermentation into a living factory known inside and out, scrutinized in all its molecular splendor, and manipulated gene by gene to perform astounding feats of biotechnology. This inspiring microbe, the sugar fungus *Saccharomyces cerevisiae*, is a secular deity, something to be revered as much as the warmth of the sun.

NOTES

Chapter 1

1. The quotes are taken from the abridged edition of the dictionary: S. Johnson, *A Dictionary of the English Language* (London: J. Knapton, et al., 1756).

2. J. A. Barnett and L. Barnett, *Yeast Research: A Historical Overview* (Washington, DC: ASM Press, 2011).

3. Franz Meyen described three species of *Saccharomyces* in 1838: *Saccharomyces cerevisiae, Saccharomyces pomorum*, and *Saccharomyces vini*. Historical details are provided in Barnett and Barnett (n. 2).

4. The most important innovation in microscopy in the 1830s was the development of achromatic lenses that removed the rainbows of colored light associated with uncorrected optics.

5. The German organic chemist Justus von Liebig never accepted the microbial explanation of fermentation, and the French naturalist Félix Pouchet refused to accept the related refutation of spontaneous generation by Pasteur.

6. Bread raised with yeast became a staple in the Roman Republic. P. Faas, *Around the Roman Table: Food and Feasting in Ancient Rome* (Chicago: University of Chicago Press, 1994); Pliny, *Natural History*, Books 17–19, Loeb Classical Library, translated by H. Rackham (Cambridge, MA: Harvard University Press, 1950).

7. A. Schriver et al., *Chemical Physics* 334, 128–37 (2007); H. Karttunen et al. (eds.), *Fundamental Astronomy*, 3rd edition (Berlin, Heidelberg: Springer, 1996).

8. The phenomenon is named after Herbert Grace Crabtree, whose research involved the metabolic competition between cancerous and non-cancerous cells that is crucial in tumor development: H. G. Crabtree, *Biochemical Journal* 23, 536–45 (1929); R. H. De Deken, *Journal of General Microbiology* 44, 149–56 (1966); R. Diaz-Ruiz, M. Rigoulet, and A. Devin, *Biochimica et Biophysica Acta* 1807, 568–76 (2011); T. Pfeiffer and A. Morley, *Frontiers in Molecular Biosciences* 1, 1–6 (2014).

9. K. H. Wolfe and D. C. Shields, *Nature* 387, 708–13 (1997). Genomic analysis by M. Marcet-Houben and T. Gabaldón in *PLoS Biology* 13(8), e1002220 (2015), suggests that *Saccharomyces cerevisiae* arose from mating between two different yeast species. In one of the possible scenarios, the resulting hybrid lived for millions of years, forming daughter cells by budding, but not undergoing sexual reproduction. The reason for this celibate interlude is that chromosomes from the different parents were sufficiently distinct that they could not have formed pairs (which is essential for sex). A comparable pairing problem is the reason that mules, which are hybrids between horses and donkeys (which have different numbers of chromosomes), are sterile. To restore the sexuality that we see in yeast today, the whole of the hybrid genome must therefore have been duplicated to provide every chromosome with a partner. The succeeding generation of yeast would then have been equipped with twice the DNA content of either of the original parents.

10. P. F. Cliften et al., *Genetics* 172, 863–72 (2006); K. H. Wolfe, *PLoS Biology* 13(8), e1002221 (2015). Duplication of the yeast genome equipped the new yeast strain with around 10,400 genes that encoded proteins. The subsequent loss of 85–90 percent of the duplicate genes trimmed this number back to 5,770 genes.

11. J. Piškur et al., *Trends in Genetics* 22, 183–6 (2006); S. Dashko et al., *FEMS Yeast Research* 14, 826–32 (2014).

12. K. Vanneste, S. Maere, and Y. van de Peer, *Philosophical Transactions of the Royal Society B* 369, 20130353 (2014).

13. Ice flavored with fruit juices was a favorite of Emperor Nero in the first century AD. The first dairy ice cream was produced in China during the Tang Dynasty (618–907 AD), by combining ice with buffalo milk, flour, and camphor.

14. Considering volume rather than width, seventy bacteria could fit inside a typical yeast cell. This is the same size ratio as elephant to human, and blue whale to elephant.

15. In the 1830s, Charles Cagnaird-Latour was the first investigator to see the scars on the yeast surface left by the separation of buds (*cicatrice[s]*), and the navel (*marque ombilicale*) on each cell that marked her separation, as a bud, from her mother. See Barnett and Barnett (n. 2).

16. According to the *OED*, Al Capp's cartoon shmoo, "is small and round, and ready to fulfil immediately any of man's material wants." The same definition might apply to yeast. In yeast biology, shmoo refers to the projection used in mating rather the whole cell.

17. T. Replansky et al., *Trends in Ecology and Evolution* 23, 294–501 (2008).

18. C. F. Kurat et al., *Journal of Biological Chemistry* 281, 491–500 (2006).
19. C. P. Kurtzman, J. W. Fell, and T. Boekhout, eds., *The Yeasts: A Taxonomic Study*, 5th edition (Amsterdam: Elsevier, 2011). The descriptions in this chapter also draw upon a second, competitor book on yeast taxonomy, namely, J. A. Barnett, R. W. Payne, and D. Yarrow, *Yeasts: Characteristics and Identification*, 3rd edition (Cambridge: Cambridge University Press, 2000).
20. The Phylum Ascomycota is subdivided into three Subphyla: Pezizomycotina, Saccharomycotina, and Taphrinomycotina. *Saccharomyces* is classified within the Saccharomycotina and *Schizosaccharomyces* fits in the Taphrinomycotina. Subphyla are big taxonomic groups. Other yeasts belong to a separate phylum, the Basidiomycota, which includes the fungi that fruit as mushrooms. The evolutionary separation of these groups of fungi is immense. Applying the taxonomic hierarchy as a very dubious measure of evolutionary distances, we would conclude that *Saccharomyces* and *Schizosaccharomyces* are as different as sea squirts (Subphylum Tunicata) and sea lions (Subphylum Vertebrata). Using the same qualitative comparison, basidiomycete and ascomycete yeasts are as far apart as sea urchins (Phylum Echinodermata) and sea lions (Phylum Chordata). An overview of fungal taxonomy is provided in N. P. Money, *Fungi: A Very Short Introduction* (Oxford: Oxford University Press, 2016).
21. R. Thaxter, *Memoirs of the American Academy of Arts and Sciences* 12, 187–249 (1896); 13, 217–649 (1908); 14, 309–426 (1924); 15, 427–580 (1926); 16, 1–435 (1931). Thaxter's work concerned a single class of tiny ascomycetes called the Laboulbeniomycetes. The glorious illustrations of other ascomycetes by Charles Tulasne (1816–84) are discussed in N. P. Money, *Mr. Bloomfield's Orchard: The Mysterious World of Mushrooms, Molds, and Mycologists* (New York: Oxford University Press, 2002).
22. K. Nasmyth, *Cell* 107, 689–701 (2001).

Chapter 2

1. F. Wiens et al., *PNAS* 105, 10426–31 (2008).
2. The maximum alcohol concentration of the palm nectar is 3.8 percent, comparable with many American light beers and draught bitter served in British pubs. Calculations by Wiens et al. (n. 1) suggest that a woman of average weight would have to consume 115 ml of pure alcohol to match the alcohol intake of a treeshrew.
3. Some work suggests that yeasts that grow in nectar have lost their odors, in instances of adaptive anonymity, to avoid detection by pollinators

that might otherwise eschew the flowers that they inhabit. M. Moritz, A. M. Yurkov, and D. Begerow, *bioRxiv* doi: https://doi.org/10.1101/088179.

4. J. Stökl et al., *Current Biology* 20, 1846–52 (2010).

5. D. N. Orbach et al., *PLoS ONE* 5(2), e8993 (2010); F. Sánchez et al., *Behavioural Processes* 84, 555–8 (2010). Experiments on the effects of alcohol on other animals include a study of an Israeli bird called the yellow-vented bulbul: S. Mazeh et al., *Behavioural Processes* 77, 369–75 (2008). The bulbul is attracted to food supplemented with enough alcohol to match the strength of fermenting fruit, but is put off by food with an alcohol content comparable to light beer.

6. Sánchez et al. (n. 5)

7. L.-A. Delegorgue, *Travels in Southern Africa*, translated by F. Webb (Durban: Killie Campbell Africana Library; Pietermaritzburg: University of Natal Press, 1990), p. 275.

8. S. Morris, D. Humphreys, and D. Reynolds, *Physiological and Biochemical Zoology* 79, 363–9 (2006).

9. L. P. Winfrey, *The Unforgettable Elephant* (New York: Walker and Co., 1980).

10. K. J. Hockings et al., *Royal Society Open Science* 2, 150150 (2015).

11. R. Dudley, *Quarterly Review of Biology* 75, 3–15 (2000); R. Dudley, *Integrative and Comparative Biology* 44, 315–23 (2004); R. Dudley, *The Drunken Monkey: Why We Drink and Abuse Alcohol* (Berkeley, CA: University of California Press, 2014).

12. T. L. Wall, S. E. Luczak, and S. Killer-Sturmhöfel, *Alcohol Research: Current Reviews* 31, 59–68 (2016).

13. M. A. Carrigan et al., *PNAS* 112, 458–63 (2015).

14. N. J. Dominy, *PNAS* 112, 3080309 (2015).

15. J. Mercader, *Science* 326, 1680–3 (2009); B. Hayden, N. Canuel, and J. Shanse, *Journal of Archaeological Methods and Theory* 20, 102–50 (2013).

16. A. Tutuola, *The Palm-Wine Drinkard* (London: Faber and Faber, 1952).

17. M. Stringini et al., *Food Microbiology* 26, 415–20 (2009).

18. S. Harmand et al., *Nature* 521, 310–15 (2015).

19. C. Lévi-Strauss, *From Honey to Ashes: Introduction to a Science of Mythology*, vol. 2 (New York: Harper and Row, 1973).

20. P. D. Sniegowski, P. G. Dombrowski, and E. Fingerman, *FEMS Yeast Research* 1, 299–306 (2002); C. T. Hittinger, *Trends in Genetics* 29, 309–17 (2013).

21. Q.-M. Wang et al., *Molecular Ecology* 21, 5404–17 (2012).

22. J.-L. Legras et al., *Molecular Ecology* 16, 2091–102 (2007); R. Tofalo et al., *Frontiers in Microbiology* 4, 1–13 (2013). The saga of yeast migration is complicated by the discovery of wild yeasts in Mediterranean forests by P. Almeida

et al., *Molecular Ecology* 24, 5412–27 (2015). Wine yeasts may have been domesticated from these fungi growing on oaks in Portugal and other countries in southern Europe. The origins of these wild strains are unclear.

23. Legras (n. 22).

24. The taxonomy of *Saccharomyces* is very complicated. The lager yeast, *Saccharomyces pastorianus*, which used to be known as *Saccharomyces carlsbergensis*, is a hybrid between *Saccharomyces cerevisiae* and *Saccharomyces eubayanus*. Similarly, Belgian-style beers are fermented by hybrid strains of *Saccharomyces cerevisiae* × *Saccharomyces kudriavzevii*. Thirdly, *Saccharomyces bayanus*, used in the fermentation of champagne, may have evolved following the hybridization between *Saccharomyces uvarum* and yeast strains formed by earlier hybridizations between *Saccharomyces cerevisiae* × *Saccharomyces eubayanus*. The nature of these domesticated hybrids has led most yeast biologists to conclude that *Saccharomyces pastorianus* and *Saccharomyces bayanus* are not distinct species. B. Dujon, *Trends in Genetics* 22, 375–87 (2006); D. Libkind et al., *PNAS* 108, 14539–44 (2011); C. T. Hittinger, *Trends in Genetics* 29, 309–17 (2013); J. Wendland, *Eukaryotic Cell* 13, 1256–65 (2014); S. Marsit and S. Dequin, *FEMS Yeast Research* 15, fov067 (2015); D. Peris, et al. *PLoS Genetics* 12(7): e1006155 (2016).

25. B. Gallone et al., *Cell* 166, 1397–410 (2016).

26. N. P. Money, *Fungi: A Very Short Introduction* (Oxford: Oxford University Press, 2016).

27. R. Mortimer and M. Polsinelli, *Research in Microbiology* 150, 199–204 (1999).

28. E. Ocón et al., *Food Control* 34, 261–7 (2013).

29. I. Stefanini et al., *PNAS* 109, 13398–403 (2012).

30. <https://ec.europa.eu/agriculture/wine/statistics_en>.

31. R. Tofalo et al., *Food Microbiology* 39, 7–12 (2014).

32. I. Stefanini et al., *PNAS* 113, 2247–51 (2016); M. Blackwell and C. P. Kurtzman, *PNAS* 113, 1971–3 (2016).

33. Coluccio et al., *PLoS ONE* 3(6): e2873 (2008); *Saccharomyces* ascospores can also overwinter in forest soils: S. J. Knight and M. R. Goddard, *FEMS Yeast Research* 16, fov102 (2016).

34. J. F. Christiaens et al., *Cell Reports* 9, 425–32 (2014).

35. N. H. Scheidler et al., *Scientific Reports* 5, 14059 (2015).

36. A. V. Devineni and U. Heberlein, *Annual Review of Neuroscience* 36, 121–38 (2013). This authoritative review article provides a nice overview of the research on fruit flies as models for alcohol research. Crayfish are also used for investigations on mechanisms of alcohol intoxication: M. E. Swierzbinski, A. R. Lazarchik, and J. Herberholz, *Journal of Experimental Biology* 220, 1516–23 (2017).

37. A. V. Devineni and U. Heberlein, *Current Biology* 19, 2126–32 (2009).
38. H. G. Lee et al., *PLoS ONE* 3(1), e1391 (2008).
39. G. Shohat-Ophir et al., *Science* 335, 1351–5 (2012).
40. K. D. McClure, R. L. French, and U. Heberlein, *Disease Models and Mechanisms* 4, 335–46 (2011).
41. The purported relationship between oxytocin and trust has been questioned by recent research: G. Nave, C. Camerer, and M. McCullough, *Perspectives in Psychological Science* 10, 772–89 (2015).

Chapter 3

1. A. Revedin et al., *PNAS* 107, 18815–19 (2010).
2. W. Rubel, *Bread: A Global History* (London: Reaktion Books, 2011).
3. P. Faas, *Around the Roman Table: Food and Feasting in Ancient Rome* (Chicago: University of Chicago Press, 1994).
4. Juvenal, *The Sixteen Satires*, translated by P. Green, Satire V, lines 79–81 (London: The Folio Society, 2014), p. 88; H. Morgan, *Bakers and the Baking Trade in the Roman Empire: Social and Political Responses from the Principate to Late Antiquity* (Master's Thesis, Pembroke College, University of Oxford, 2015).
5. Pliny, *Natural History*, Books 17–19, Loeb Classical Library, translated by H. Rackham (Cambridge, MA: Harvard University Press, 1950).
6. J. R. Clarke, *The Houses of Roman Italy, 100 B.C.–A.D. 250: Ritual, Space, and Decoration, Part 250* (Berkeley, CA: University of California Press, 1991); A. Cooley, *The Cambridge Manual of Latin Epigraphy* (Cambridge: Cambridge University Press, 2012).
7. P. W. Hammond, *Food and Feast in Medieval England* (Gloucestershire, United Kingdom: Alan Sutton, 1993); J. L. Singman, *Daily Life in Medieval Europe* (Westport, CT: Greenwood Press, 1999).
8. E. Buehler, *Bread Science: The Chemistry and Craft of Making Bread* (Hillsborough, NC: Two Blue Books, 2006).
9. D. F. Good, *The Economic Rise of the Habsburg Empire, 1750–1914* (Berkeley, CA: University of California Press, 1984); M. Roehr, in *History of Modern Biotechnology I*, edited by A. Fiechter (Amsterdam: Springer, 2000), 127–8.
10. P. Debré, *Louis Pasteur*, translated by E. Forster (Baltimore, MD: Johns Hopkins University Press, 1998).
11. E. N. Horsford, *Report on Vienna Bread* (Washington, DC: Government Printing Office, 1875).
12. A. de Tocqueville, *Democracy in America and Two Essays on America*, translated by G. E. Bevan (London: Penguin, 2003) 526.

13. P. C. Klieger, *Images of America: The Fleischmann Yeast Family* (Charleston, SC: Arcadia, 2004).

14. A test tube containing a dried culture of the original yeast strain used by Fleischmann would be a very valuable historical artifact. A preserved culture of the strain of *Penicillium notatum* that produced the antibiotic discovered by Alexander Fleming in 1928, was sold at auction in 2016 for US$46,000: E. Blakemore, <http://www.smithsonian.com>, December 9, 2016.

15. M. Morgan, *Over-the-Rhine: When Beer Was King* (Charleston, SC: The History Press, 2010); S. Stephens, *Images of America: Cincinnati's Brewing History* (Charleston, SC: Arcadia, 2010).

16. <http://www.breadworld.com/history>.

17. *Yeast Market: Global Trends and Forecast to 2020*, Markets and Markets Report Code FB 2233 <http://www.marketsandmarkets.com>.

18. A. Bekatorou et al., *Food Technology and Biotechnology* 44, 407–15 (2006); R. Gómez-Pastor et al., in *Biomass: Detection, Production and Usage*, edited by D. Matovic <http://www.intechopen.com> (2011), 201–22.

19. H. Takagi and J. Shima, in *Stress Biology of Yeasts and Fungi: Applications for Industrial Brewing and Fermentation*, edited by H. Takagi and H. Kitagaki (Tokyo: Springer Japan, 2015), 23–42.

20. S. P. Cauvain and L. S Young, editors, *The Chorleywood Bread Process* (Cambridge: Woodhead Publishing, 2006).

21. <http://www.chopin.fr/en/produits/3-alveograph.html>.

22. H. G. Müller, *Transactions of the British Mycological Society* 41, 341–64 (1958). Muller utilized his knowledge of baking in a brief book on the origins and development of the industry in Europe: H. G. Muller, *Baking and Bakeries* (Oxford: Shire Publications, 1986).

23. <https://www.wonderbread.com/products/classic-white/>.

24. A. Whitley, *Bread Matters: The State of Modern Bread and Definitive Guide to Baking Your Own* (London: 4th Estate, 2009).

25. L. De Vuyst and P. Neysens, *Trends in Food Science and Technology* 16, 43–56 (2005).

26. R. F. Schwan and A. E. Wheals, *Critical Reviews in Food Science and Nutrition* 44, 205–21 (2004); D. S. Nielsen et al., in *Chocolate in Health and Nutrition*, edited by R. R. Watson, V. R. Preedy, and S. Zibadi (New York: Humana, 2013), 39–60; V. T. Ho, J. Zhao, and G. Fleet, *International Journal of Food Microbiology* 174, 72–87 (2014).

27. H.-M. Daniel et al., *FEMS Yeast Research* 9, 774–83 (2009).

28. J. A. Barnett, R. W. Payne, and D. Yarrow, *Yeasts: Characteristics and Identification*, 3rd edition (Cambridge: Cambridge University Press, 2000);

C. P. Kurtzman, J. W. Fell, and T. Boekhout, eds., *The Yeasts: A Taxonomic Study*, 5th edition (Amsterdam: Elsevier, 2011).

29. Z. Papalexandratou et al., *Food Microbiology* 35, 73–85 (2013).

30. C. L. Ludlow et al., *Current Biology* 26, 1–7 (2016).

31. C. Price, *Vitamania: Our Obsessive Quest for Nutritional Perfection* (New York: Penguin Press, 2015); C. Price, *Distillations Magazine* Fall Issue, 17–23 (2015); Klieger (n. 13).

32. Klieger (n. 13).

33. I. Goldberg, *Single Cell Protein* (Berlin, Heidelberg: Springer, 2013).

34. M. Buggein, *Slave Labor in Nazi Concentration Camps* (Oxford: Oxford University Press, 2014).

35. J. A. Barnett, *Yeast* 20, 509–43 (2003).

36. J. L. Marz, *A Revolution in Biotechnology* (Cambridge: Cambridge University Press, 1989).

37. T. Shabad, *New York Times*, November 10, 1973; K. Wolf, *Nonconventional Yeasts in Biotechnology: A Handbook* (Berlin, New York: Springer, 1996); M. L. Rabinovich, *Cellulose Chemistry and Technology* 44, 173–86 (2010).

38. <https://www.cia.gov/library/readingroom/document/0000498552>.

39. *Yeast Market* (n. 17).

40. R. J. Gruninger et al., *FEMS Microbial Ecology* 90, 1–17 (2014).

41. J. Callister, *The Man Who Invented Vegemite: The Story Behind an Australian Icon* (Murdoch Books, 2012); <http://www.ilovemarmite.com/default.asp>.

42. J. Lewis, *The Jerusalem Post*, June 18, 2015.

Chapter 4

1. F. M. Klis, C. G. de Koster, and S. Brul, *Eukaryotic Cell* 13, 2–9 (2014); <http://bionumbers.hms.harvard.edu/default.aspx>.

2. D. S. Goodsell, *The Machinery of Life*, 2nd edition (New York: Copernicus/Springer, 2009).

3. D. Araiza-Olivera et al., *The FEBS Journal* 280, 3887–905 (2013).

4. Vesicle formation and trafficking through the cytoplasm are fundamental to the life of all eukaryotic cells. The elucidation of this endomembrane transport process involved elegant experiments on yeast by Randy Schekman, and on mammalian cells by James Rothman and Thomas Südhof. These scientists shared the 2013 Nobel Prize in Physiology or Medicine: I. Mellman and S. D. Emr, *Journal of Cell Biology* 203, 559–61 (2013). Yoshinori Ohsumi won the Nobel Prize in 2016 for his work on yeast mutants that revealed how the endomembrane system is reorganized during the continuous turnover of proteins and breakdown of organelles

within the cell. This mechanism is called autophagy and its operation in human cells plays a critical role in the progression of many serious diseases: F. Reggiori and D. J. Klionsky, *Genetics* 194, 341–61 (2013).

5. M. Osumi, *Journal of Electron Microscopy* 61, 343–65 (2012).
6. M. C. Gustin et al., *Science* 233, 1195–7 (1986).
7. A. B. G. Goffeau et al., *Science* 274, 546–67 (1996).
8. S. Wilkening et al., *BMC Genomics* 14, 90 (2013).
9. B. Dujon, *FEMS Yeast Research* 15, fov047 (2015).
10. J. A. Barnett, *Yeast* 24, 799–845 (2007).
11. C. C. Lindegren et al., *Nature* 183, 800–2 (1959).
12. Carl Lindegren detailed his bizarre views on inheritance in a strange book titled, *The Cold War Biology* (Ann Arbor, MI: Planarian Press, 1966); J. Sapp, *Beyond the Gene: Cytoplasmic Inheritance and the Struggle for Authority in Genetics* (New York: Oxford University Press, 1987).
13. D. Botstein and G. R. Fink, *Genetics* 189, 695–704 (2011); S. R. Engel and J. M. Cherry, *Database* 2013, bat012 (2013); <http://www.yeastgenome.org>.
14. H. Feldmann, *Yeast: Molecular and Cell Biology*, 2nd edition (Weinheim, Germany: Wiley-Blackwell, 2012).
15. C. L. Ludlow, *Nature Methods* 10, 671–5 (2013).
16. G. Giaever and C. Nislow, *Genetics* 197, 451–65 (2014); <http://www-sequence.stanford.edu/group/yeast_deletion_project/project_desc.html>.
17. P. F. Cliften et al., *Genetics* 172, 863–72 (2006); K. H. Wolfe, *PLoS Biology* 13(8), e1002221 (2015).
18. M. Koegl and P. Uetz, *Briefings in Functional Genomics and Proteomics* 6, 302–12 (2008); Y. C. Chen et al., *Nature Methods* 7, 667–8 (2010); Feldmann (n. 14).
19. A. H. Y. Tong and C. Boone, *Methods in Microbiology* 36, 369–86, 706–7 (2007).
20. <http://www.singerinstruments.com>.
21. Giaever and Nislow (n. 16).
22. L. Peña-Castillo and T. R. Hughes, *Genetics* 176, 7–14 (2007).
23. M. Schlackow and M. Gullerova, *Biochemical Society Transactions* 41, 1654–9 (2013).
24. *Pocket Oxford English Dictionary*, 11th edition, edited by M. White (Oxford: Oxford University Press, 2013).
25. F. M. Doolittle, *PNAS* 110, 5294–300 (2013).
26. A. F. Palazzo and T. R. Gregory, *PLoS Genetics* 10(5), e1004351 (2014).
27. C. J. Gimeno et al., *Cell* 68, 1077–90 (1992); J. M. Gancedo, *FEMS Microbiology Reviews* 25, 107–23 (2001).
28. W. C. Ratcliff et al., *Nature Communications* 6, 6102 (2015).

29. Although the mutated genes in Great Danes that give them such long legs are transferred by sex, the principle is exactly the same as the asexual phenomenon of snowflaking in yeast. Great Danes tend to suffer from digestive problems, heart disease, and hip dysplasia, but these disabilities are considered secondary to the desirable features of the stature and gentle disposition of the breed.

30. W. C. Ratcliff et al., *PNAS* 109, 1595–600 (2012); K. Voordeckers and K. J. Verstrepen, *Current Opinion in Microbiology* 28, 1–9 (2015). Related studies have shown that yeast cells tend to clump together when sucrose is scarce: J. H. Koschwanez, K. R. Foster, and A. W. Murray, *PLoS Biology* 9(8), e1001122 (2011). This natural resistance to separation after cell division allows the participants to cooperate in sucrose digestion and concentrate glucose and fructose within the clump. This kind of cooperative behavior may have been a feature in the evolution of multicellularity.

31. T. T. Hoffmeyer and P. Burkhardt, *Current Opinion in Genetics and Development* 39, 42–7 (2016).

32. A. H. Kachroo et al., *Science* 348, 921–5 (2015).

33. N. Annaluru et al., *Science* 344, 55–8 (2014); S. M. Richardson et al., *Science* 355, 1040–4 (2017); <http://syntheticyeast.org/sc2o/introduction/>.

34. R. K. Mortimer and J. R. Johnston, *Genetics* 113, 35–43 (1986).

35. <http://syntheticyeast.org/sc2-o/ethics-governance/>.

36. E. Darwin, *Zoonomia; or the Laws of Organic Life* (London: Printed for J. Johnson, 1794–6).

37. A. Pross, *What is Life? How Chemistry Becomes Biology*, 2nd edition (Oxford: Oxford University Press, 2016); N. Lane, *The Vital Question: Energy, Evolution, and the Origins of Complex Life* (New York: W. W. Norton & Company, 2015).

Chapter 5

1. <https://www.fhwa.dot.gov/environment/climate_change/adaptation/publications/climate_effects/effects03.cfm>.

2. Data compiled by the University of Iowa for 2013–14 showed that 39 percent of 87.5 million acres (554,099 square kilometers) of harvested corn was dedicated to bioethanol production. This corresponds to a crop area of 138,098 square kilometers: <https://www.extension.iastate.edu/agdm/crops/outlook/cornbalancesheet.pdf>.

3. Report from the Institute for Energy Research (IER): <http://institute forenergyresearch.org/topics/encyclopedia/biomass/>.

4. C. K. Wright and M. C. Wimberly, *PNAS* 110, 4134–9 (2013).

5. A. E. Farrell et al., *Science* 311, 506–8 (2006); Z. Wang, J. Dunn, and M. Wang, *Updates to the Corn Ethanol Pathway and Development of an Integrated Corn and Corn Stover Ethanol Pathway in the GREET Model* (Argonne, IL: Argonne National Laboratory, 2014) and associated reports available at <https://greet.es.anl.gov/publications>.

6. D. D. Smith, *Journal of the Iowa Academy of Sciences* 105, 94–108 (1998).

7. L. C. Basso, et al., *FEMS Yeast Research* 8, 1155–63 (2008).

8. B. E. Della-Bianca et al., *Applied Microbiology and Biotechnology* 97, 979–91 (2013).

9. J. Shima and T. Nakamura, in *Stress Biology of Yeasts and Fungi*, edited by H. Takagi and H. Kitagaki (Tokyo: Springer Japan, 2015), 93–106.

10. Apprentices to chimney sweeps were, essentially, owned by their masters. Boys as young as six suffered blisters and serious burns as they shunted themselves inside hot chimneys, developed skeletal deformities, and, if they survived childhood, developed testicular cancer from their exposure to coal tar. The occupation is associated with the Victorian era, but became common in the late 1700s.

11. L. Caspeta et al., *Science* 346, 75–8 (2014); C. Cheng and K. C. Katy, *Science* 346, 35–6 (2014).

12. D. Stanley et al., *Journal of Applied Microbiology* 109, 13–24 (2010); F. Lam et al., *Science* 346, 71–5 (2014).

13. J. E. DiCarlo et al., *Nucleic Acids Research* 41, 4336–43 (2013); G.-C. Zhang et al., *Applied and Environmental Microbiology* 80, 7694–701 (2014); V. Stovicek, I. Borodina, and J. Forster, *Metabolic Engineering Communications* 2, 13–22 (2015).

14. R. Mans et al., *FEMS Yeast Research* 15, fov004 (2015).

15. A. C. Komor et al., *Nature* 533, 420–4 (2016).

16 J. E. DiCarlo et al., *Nature Biotechnology* 33, 1250–5 (2015); J. Perkel, *Biotechniques* 58, 223–7 (2015).

17. N. P. Money, *Fungi: A Very Short Introduction* (Oxford: Oxford University Press, 2016).

18. D. Klein-Marcuschamer et al., *Biotechnology and Bioengineering* 109, 1083–7 (2012).

19. A. J. Liska et al., *Nature Climate Change* 4, 398–401 (2014).

20. R. Yamada et al., *Biotechnology for Biofuels* 4, 1–8 (2011); Yamada et al., *AMB Express* 3, 1–7 (2013).

21. J. Nielsen et al., *Current Opinion in Biotechnology* 24, 398–404 (2013).

22. M. Kanellos, <http://www.greentechmedia.com/articles/read/can-isobutanol-replace-ethanol>, June 1, 2011; M. Fellet, *Chemical and Engineering News* 94 (37), 16–19 (2016).

23. <https://www.nobelprize.org/nobel_prizes/medicine/laureates/2015/tu-facts.html>.
24. C. J. Paddon and J. D. Keasling, *Nature Reviews Microbiology* 12, 355–67 (2014). Funding from the Gates Foundation supported a successful partnership between researchers at the University of California, a for-profit company, and a non-profit company.
25. Greater appreciation of the exigencies of life in rural Africa is evident in *Tarzan the Ape Man* that starred Johnny Weissmuller in 1932.
26. <http://www.marketsandmarkets.com/PressReleases/human-insulin.asp>.
27. B. Lam, *The Atlantic*, February 10, 2015.
28. L. E. Markowitz et al., *Pediatrics* 137, e20151968 (2016).
29. P. Stalmans et al., *The New England Journal of Medicine* 367, 606–15 (2012).
30. Nutritional yeast is a rich source of the vitamin.
31. E. Whitman, <http://www.ibtimes.com>, April 10, 2015; <http://www.drugabuse.gov/related-topics/trends-statistics>.
32. My phrase "tranquilizing spirits" was inspired by the temptation of a young lady by Comus, son of Bacchus and Circe, in John Milton's play of the same name:

> But this will cure all streight, one sip of this
> Will bathe the spirits in delight
> Beyond the bliss of dreams.
> *Comus (A Mask Presented at Ludlow Castle, 1634), 810–12*

33. W. C. DeLoache et al., *Nature Chemical Biology* 11, 465–71 (2015).
34. K. Thodey, S. Galanie, and C. D. Smolke, *Nature Chemical Biology* 10, 837–44 (2014); E. Fossati et al., *PLoS ONE* 10(4), e0124459 (2015).
35. < http://www.skunk-skunk.com>.
36. W. Davies, *BBC News* 2 October 2008, <http://news.bbc.co.uk/2/hi/middle_east/7646894.stm>.
37. Conventional pesticides seem to be more damaging to a wider range of non-target insects than the Bt toxin expressed in GM corn: J. A. Peterson, J. G. Lundgren and J. D. Harwood, *The Journal of Arachnology* 39, 1–21 (2011).
38. O. Tokareva et al., *Microbial Biotechnology* 6, 651–63 (2013).
39. <http://www.boltthreads.com>.

Chapter 6

1. L. G. Nagy et al., *Nature Communications* 5, 4471 (2014).
2. C. T. Ingold, *Transactions of the British Mycological Society* 86, 325–8 (1986).
3. N. P. Money, *Nature* 465, 1025 (2010).

4. The Latin name, *Auricularia auricula-judae*, means Jew's ear, which refers to Judas, who hanged himself from an elder tree as penance for his sin. Elder is one of the plants on which this fungus feeds and fruits. The Biblical reference and unintentional anti-Semitism are avoided in mushroom guidebooks that shorten the name to *Auricularia auricula*. See N. P. Money, *Mushrooms: A Natural and Cultural History* (London: Reaktion Books, 2017).

5. The mirror-picture-formation, or *Spiegelbilderzeugung* (in the tongue of its discoverers in the 1890s), works best when a culture dish is inverted so that the spores fall on to the underside of the lid after they are released from the yeast colony. The spores are discharged over a distance of 0.5 millimeters, but their electrostatic attraction to the plastic used to manufacture modern Petri plates also allows the fungus to paint the underside of its lids without inversion. This footnote is written in full recognition that it will be of interest to no more than three or four people.

6. The number of professional funambulists is probably similar to the number of professional mycologists.

7. A. H. R. Buller, *Researches on Fungi*, vol. 7 (Toronto: University of Toronto Press, 1950). The rules of fungal taxonomy dictate the substitution of *Sporidiobolus* for the first names of some species of *Sporobolomyces* now.

8. *Mycological Society of America Newsletter* XI, 1 (June 1960).

9. L. G. Goldsborough, "Reginald Buller: The Poet-Scientist of the Mushroom City," *Manitoba History* XLVII, 17–41 (2004). His one great poem was a limerick concerning Einstein's general theory of relativity. This was published in *Punch* in 1923:

> There was a young lady named Bright,
> Whose speed was far faster than light;
> She started one day
> In a relative way,
> And returned on the previous night.

10. A. Pringle et al., *Mycologia* 97, 866–71 (2005).

11. Ilya Ilyich Mechnikov (1845–1916) adopted the French name Élie and his surname appears as Metchnikoff and Metschnikoff. The great scientist was published as Elias Metschnikoff in his paper on the fungal infection of *Daphnia* in 1884 and this spelling was applied to the Latin name of the yeast genus when it was described formally in 1899. Metchnikoff has been preferred by biographers. The spores are described by M.-A. Lachance et al., *Canadian Journal of Microbiology* 44, 279–88 (1998).

12. M.-A. Lachance et al., *Canadian Journal of Microbiology* 22, 1756–61 (1976). There is some educated guesswork in my description of the process of

spore ejection because nobody has caught the fungus in the act of harpooning the insect gut.

13. C. P. Kurtzman, J. W. Fell, and T. Boekhout, eds., *The Yeasts: A Taxonomic Study*, 5th edition (Amsterdam: Elsevier, 2011).

14. T. D. Brock, *Milestones in Microbiology: 1546–1940* (Washington, DC: ASM Press, 1999).

15. A. I. Tauber and L. Chernyak, *Metchnikoff and the Origins of Immunology: From Metaphor to Theory* (New York: Oxford University Press, 1991); L. Vikhanski, *Immunity: How Elie Metchnikoff Changed the Course of Modern Medicine* (Chicago: Chicago Review Press, 2016).

16. S. Kaufmann, *Nature Immunology* 9, 705–12 (2008). Metchnikoff described phagocytosis in starfish larvae that he pricked with rose thorns. The experiments with *Daphnia* showed that the same inflammatory cellular response took place in a simple animal infected by a microorganism.

17. L. Carroll, *Through the Looking-Glass, and What Alice Found There* (London: Macmillan, 1871).

18. S. R. Hall et al., *American Naturalist* 174, 149–62 (2009).

19. M.-A. Lachance and W.-M. Pang, *Yeast* 13, 225–32 (1997).

20. M. J. Schmitt and F. Breinig, *Nature Microbiology Reviews* 4, 212–21 (2006).

21. M. J. Schmitt and F. Breinig, *FEMS Microbiology Reviews* 26, 257–76 (2002); M. F. Perez et al., *PLoS ONE* 11(10), e0165590 (2016).

22. M. F. Madelin and A. Feast, *Transactions of the British Mycological Society* 79, 331–5 (1982). Mike Madelin (1931–2007) was a professor at the University of Bristol.

23. As devotees of Darwin know, Robert McCormick was appointed as the ship's surgeon at the beginning of the second voyage on HMS *Beagle* to survey the coasts of South America. It was customary for the surgeon to work as the ship's naturalist and McCormick looked forward to the opportunities for collecting plants and animals during the voyage. Darwin was engaged as a companion to Captain Robert Fitzroy and self-financed naturalist. Frustrated by the presence of Darwin as a second naturalist on the same ship, McCormick left the expedition in its fourth month. J. W. Gruber, *The British Journal for the History of Science* 4, 266–82 (1969); H. L. Burstyn, *The British Journal for the History of Science* 8, 62–9 (1975).

24. J. L. F. Kock et al., *South African Journal of Science* 100, 237–40 (2004).

25. Strains of *Saccharomyces* can be nudged toward hyphal behavior by exposing the fungus to stressful environmental conditions that yield chains of cells (see Chapter 4). This has been called pseudo-hyphal growth. *Dipodascus* goes much further than the sugar fungus, elaborating three-dimensional colonies of branching hyphae.

26. L. R. Tulasne and C. Tulasne, *Selecta Fungorum Carpologia*, 3 volumes (Paris: Imperatoris Jussu, In Imperiali Typographeo Excudebatur, 1861–5). Carpology is the branch of botany concerned with the study of fruits and seeds. According to modern terminology, which specifies that fruits and seeds are produced by plants, we say that the Tulasne brothers described the fruit bodies and spores of fungi.

27. N. P. Money, *Fungal Biology* 117, 463–5 (2013).

28. H. Nilsson et al., *Evolutionary Bioinformatics Online* 4, 193–201 (2008); R. Blaalid et al., *Molecular Ecology Resources* 13, 218–24 (2013).

29. Kurtzman, Fell, and Boekhout (n. 13).

30. W. T. Starmer and M.-A. Lachance, *Yeast Ecology*, in C. P. Kurtzman, J. W. Fell, and T. Boekhout, *The Yeasts: A Taxonomic Study*, 5th edition (Amsterdam: Springer, 2011), 88–107.

31. J. A. Barnett, R. W. Payne, and D. Yarrow, *Yeasts: Characteristics and Identification*, 3rd edition (Cambridge: Cambridge University Press, 2000).

32. R. Richle and H. J. Scholer, *Pathologia et Microbiologia* 24, 783–93 (1961); C. H. Zierdt et al., *Antonie Van Leeuwenhoek* 54, 357–66 (1988).

33. S. N. Kutty and R. Philip, *Yeast* 25, 465–83 (2008).

34. D. Bass et al., *Proceedings of the Royal Society B* 274, 3069–77 (2007).

35. K. Takishita et al., *Extremophiles* 10, 165–9 (2006).

36. N. P. Money, *The Amoeba in the Room: Lives of the Microbes* (Oxford: Oxford University Press, 2014).

37. K.-S. Shin et al., *International Journal of Systematic and Evolutionary Microbiology* 51, 2167–70 (2001). A filamentous fungus and a red alga are the most heat-tolerant eukaryotes, capable of growing at 55°C.

38. Starmer and Lachance (n. 30).

39. F. Branda et al., *FEMS Microbiology Ecology* 72, 354–69 (2010).

40. J. D. Castello and S. O. Rogers, eds., *Life in Ancient Ice* (Princeton, NJ: Princeton University Press, 2005).

41. L. Selbmann et al., *Fungal Biology* 118, 61–71 (2014).

42. M. N. Babič, et al., *Fungal Biology* 119, 95–113 (2015).

43. C. W. Bruch, in *Airborne Microbes* (Society for General Microbiology Symposium no. 17), edited by P. H. Gregory and J. L. Monteith, (Cambridge: Cambridge University Press, 1967), 345–73.

44. E. Ejdys, J. Michalak, and K. M. Szewczyk, *Acta Mycologica* 44, 97–107 (2009); R. I. Adams et al., *The ISME Journal* 7, 1262–73 (2013); A. J. Prussin and L. C. Marr, *Microbiome* 3, 78 (2015); B. Hansen et al., *Environmental Science: Processes & Impacts* 18, 713–24 (2016).

45. The original calculation of the number of airborne spores appears in Money (n. 36). To get from 10^{23} spores to an African surface area,

assume that each sphere has a diameter of ten millionths of one meter (ten micrometers), and calculate its surface area using the equation $4\pi r^2$.

46. M. O. Hassett, M. W. F. Fischer, and N. P. Money, *PLoS ONE* 10(10), e0140407 (2015).
47. C. S. Hoffman, V. Wood, and P. A. Fantes, *Genetics* 201, 403–23 (2015).
48. K. Nasmyth, *Cell* 107, 689–701 (2001).
49. V. Wood, *Nature* 415, 871–80 (2002).
50. J. M. Misihairabgwi et al., *African Journal of Microbiological Research* 9, 549–56 (2015).
51. <http://www.pombase.org>.

Chapter 7

1. P. Muñoz et al., *Clinical Infectious Diseases* 40, 1625–34 (2005); A. Enache-Angoulvant and C. Hennequin, *Clinical Infectious Diseases* 41, 1559–68 (2005); R. Pérez-Torrado and A. Querol, *Frontiers in Microbiology* 6, 1522 (2015).
2. And his brother with a faulty heart valve would have collapsed one day in a golden wheat field from no discoverable cause: *pallida Mors aequo pulsat pede*, pale death knocks with impartial foot, *pauperum tabernas regumque turris*, on the poor man's cottage and the rich man's castle. Horace, *Odes and Epodes*, Odes I.4, lines 13–14, Loeb Classical Library, translated by N. Rudd (Cambridge, MA: Harvard University Press, 2004).
3. P. K. Strope et al., *Genome Research* 25, 1–13 (2015). Subtle differences between the genomes of clinical versus non-clinical strains were found in this study, but the way that these might allow the clinical strains to flourish in the bloodstream is unclear.
4. D. P. Jensen and D. L. Smith, *Archives of Internal Medicine* 136, 332–3 (1976); K. S. C. Fung et al., *Scandinavian Journal of Infectious Diseases* 28, 83–5 (1996).
5. Unlike ordinary yeast, *Saccharomyces boulardii* grows well at human body temperature and under quite acidic conditions. These differences, along with the distinctive genetic signature of the yeast, encouraged some yeast biologists to classify this probiotic microbe as a distinct species. The problem with this assessment is that the genetic variation among strains of *Saccharomyces cerevisiae* used for brewing and baking is much greater than the differences between *boulardii* and any strains of *cerevisiae* selected for comparison: L. C. Edwards-Ingram et al., *Genome Research* 14, 1043–51 (2004). If anyone wishes to consult the original research, do not make a typo in your web search, as I did, and omit the first *e* of *Genome*.
6. Colonel Kurtz was played by Marlon Brando in the Francis Ford Coppola movie *Apocalypse Now* (1979), which transferred Joseph Conrad's 1899

novella, *Heart of Darkness*, set in King Leopold's Congo, to South East Asia. It seems unlikely that my screenplay about Henri Boulard, titled, *Zut alors! Ou est mon pot de chamber?* will be made into a movie.

7. A balanced view of the probiotic industry is provided by L. E. Miller, *Journal of Dietary Supplements* 12, 261–4 (2015).

8. D. Czerucka, T. Piche, and P. Rampal, *Alimentary Pharmacology and Therapeutics* 26, 767–78 (2007); T. Kelesidis, *Therapeutic Advances in Gastroenterology* 5, 112–25 (2012); J.-P. Buts, *Digestive Diseases and Sciences* 54, 15–18 (2009).

9. M. I. Moré and A. Swidsinski, *Clinical and Experimental Gastroenterology* 8, 237–55 (2015).

10. L. Vikhanski, <http://www.smithsonianmag.com/science-nature/science-lecture-accidentally-sparked-global-craze-yogurt-180958700/> (April 11, 2016).

11. <https://www.ftc.gov/news-events/press-releases/2010/12/dannon-agrees-drop-exaggerated-health-claims-activia-yogurt>.

12. P. D. Scanlon and J. R. Marchesi, *ISME Journal* 2, 1183–93 (2008); S. J. Ott et al., *Scandinavian Journal of Gastroenterology* 43, 831–41 (2008).

13. F. Cuskin et al., *Nature* 517, 165–9 (2015).

14. B. Cordell and J. McCarthy, *International Journal of Clinical Medicine* 4, 309–12 (2013).

15. B. K. Logan and A. W. Jones, *Medicine, Science and the Law* 40, 206–15 (2000).

16. A. Al-Awadhi et al., *Science and Justice* 44, 149–52 (2004).

17. A. Hunnisett and J. Howard, *Journal of Nutritional Medicine* 1, 33–9 (1990).

18. There is an extensive literature on anti-*Saccharomyces cerevisiae* antibodies (ASCAs) and Crohn's disease, and interested readers are referred to the Internet with the usual proviso of taking care to consult the best peer-reviewed journals as an objective guide to the topic. E. Israeli et al., *Gut* 54, 1232–6 (2005), is cited widely. The association between the antibody and Crohn's disease was first recognized in the 1980s by investigators in Scotland: J. Main et al., *British Medical Journal* 297, 1105–6 (1998).

19. F. Seibold, *Gut* 54, 1212–13 (2005). Interpretation of the presence of anti-*Saccharomyces cerevisiae* antibodies in healthy family members is complicated by the finding that the antibodies seem to be predictive of the future development of the disease. People who test positive without Crohn's disease often go on to develop symptoms later in life.

20. The popularity of gluten-free food is apparent to every shopper in the United States. There are few examples of greater hype in the food industry, but there is some evidence of gluten sensitivity among people without celiac disease: A. Fasano et al., *Gastroenterology* 148, 1195–204 (2015).

21. D. D. Karsada, *Journal of Agricultural and Food Chemistry* 61, 1155–9 (2013).
22. Again (n. 18), the literature on celiac and yeast is voluminous. One could begin by consulting the following sources and picking up the trail in the more recent literature: D. Toumi et al., *Scandinavian Journal of Gastroenterology* 42, 821–6 (2007); L. M. S. Kotze et al., *Arquivos de Gastroenterologia* 47, 242–5 (2010).
23. M. Rinaldi et al., *Clinical Reviews in Allergy and Immunology* 45, 152–61 (2013).
24. W. G. Crook, *The Yeast Connection: A Medical Breakthrough* (New York: Vintage Books, 1986).
25. J. Brisman, *Occupational and Environmental Medicine* 59, 498–502 (2002); J. Belchi-Hernandez, A. Mora-Gonzalez, and J. Iniesta-Perez, *Journal of Allergy and Clinical Immunology* 97, 131–4 (1996).
26. G. E. Packe and J. G. Ayres, *The Lancet* 326, 199–204 (July 27, 1985).
27. <http://www.aaaai.org>.
28. D. W. Cockcroft et al., *Journal of Allergy and Clinical Immunology* 72, 305–9 (1972).
29. J. Plazas et al., *AIDS* 8, 387–8 (2012).
30. V. Sharma, J. Shankar, and V. Kotamarthi, *Eye* 20, 945–6 (2006).
31. C. Beimforde et al., *Molecular Phylogenetics and Evolution* 78, 386–98 (2014).
32. C. P. Kurtzman, J. W. Fell, and T. Boekhout, eds., *The Yeasts: A Taxonomic Study*, 5th edition (Amsterdam: Elsevier, 2011).
33. P. E. Sudbery, *Nature Reviews Microbiology* 9, 737–48 (2011).
34. B. J. Kullberg and M. C. Arendrup, *The New England Journal of Medicine* 373, 1445–56 (2015).
35. <http://www.cdc.gov/fungal/diseases/candidiasis/candida-auris-alert.html>.
36. K. Seider, *Current Opinion in Microbiology* 13, 392–400 (2010).
37. A. Boroch, *The Candida Cure: Yeast, Fungus and Your Health—The 90-Day Program to Beat Candida and Restore Vibrant Health* (Studio City, CA: Quintessential Healing, 2009).
38. W. F. Nieuwenhuizen et al., *The Lancet* 361, 2152–4 (2003); M. Corouge et al., *PLoS ONE* 10(3), e0121776 (2015).
39. A. Casadevall and L.-A. Pirofski, *Nature* 516, 165–6 (2014).
40. M. A. Ghannoum et al., *PLoS Pathogens* 6(1), e1000713 (2010).
41. <http://www.cdc.gov/fungal/diseases/index.html>.
42. P. Zalar et al., *Fungal Biology* 115, 997–1007 (2011).
43. K. Findley et al., *Nature* 498, 367–70 (2013).
44. A. Velegraki et al., *PLoS Pathogens* 11(1), e1004523 (2015).
45. A. Amend, *PLoS Pathogens* 10(8), e1004277 (2014).

GLOSSARY

acetaldehyde compound (C_2H_4O) produced during alcoholic fermentation that is responsible for some hangover symptoms

aerobic with oxygen; aerobic metabolic reactions occurring in the presence of oxygen

anaerobic without oxygen; anaerobic metabolic reactions occurring in the absence of oxygen

ascomycete fungi classified in a taxonomic group called the Ascomycota, including *Saccharomyces, Aspergillus,* and other common molds, truffles, and morels

ascospore the type of spore characteristic of ascomycete fungi

ascus modified cell in which ascospores are formed

ATP abbreviation for adenosine triphosphate, a molecule that serves as a mobile carrier of chemical energy that fuels the biochemical reactions in every cell

basidiomycete fungi classified in a taxonomic group called the Basidiomycota, including mushrooms, bracket or shelf fungi, jelly fungi, and rusts and smuts that infect plants

biomass broad term for a quantity of biological material; also describes the fibrous crop waste that is used to manufacture second generation biofuels

cell wall the layer of polymers that surrounds the cells of most organisms, with the notable exception of animal cells and a variety of amoeboid microorganisms

cellulose the polysaccharide made from chains of glucose molecules, which forms much of the bulk of plant cell walls

centromere part of chromosomes that work as anchors for the molecular machinery that separates chromosomes during cell division; also the location where two copies of a chromosome produced by DNA replication are joined

choanoflagellate a single-celled organism that feeds on bacteria strained through a collar

citric acid cycle second series of reactions in aerobic respiration that release carbon dioxide and produce chemical energy in the form of ATP

codon series of three letters (nucleotides) in DNA and RNA sequences that specify the position of individual amino acids in protein synthesis; DNA codons also dictate START and STOP signals that begin and end the process of protein synthesis

dimorphism switch between two distinctive forms of growth, including the manifestation of some yeast species as single cells that form buds and as colonies of filamentous hyphae

eukaryote organism whose cells contain chromosomes housed within a nucleus; examples include yeasts and mushrooms, seaweeds, all plants, and all animals

fermentation metabolic breakdown of sugars that produces ethanol

fission yeast *Schizosaccharomyces pombe*

genome the totality of genetic material possessed by an organism

glycolysis first set of reactions in sugar metabolism that convert each molecule of glucose to two molecules of pyruvate

Golgi apparatus system of membranes within the cell that processes proteins for export

hexokinase an enzyme that participates in the breakdown of glucose or similar monosaccharide sugars

hyphae microscopic filamentous cells produced by fungi that penetrate solid food, release enzymes, and absorb nutrients through their tips

internal transcribed spacer (ITS) A DNA sequence positioned between genes that encode the RNA sequences that form ribosomes; useful for comparing the relatedness of organisms

mating type a version of one species that is incompatible with the same version in a reproductive interaction; determined by genetics; roughly equivalent to sex or gender

metagenomics analysis of DNA sequences obtained from environmental samples

mitochondria cellular organelles that produce chemical energy in the form of ATP via the citric acid cycle and associated biochemical processes

molasses black viscous syrup that is a by-product from the process of refining sugar from sugar cane and sugar beet

monophyletic groups of organisms that are not related though a common ancestral species, but have evolved along independent lines of descent from different ancestors

open reading frame (ORF) DNA sequence that is read into a protein

phagocytosis mechanism used by a cell to engulf solid particles from its surroundings

plasmid small accessory loop of DNA present in the cells of many bacteria, and some archaea and eukaryotes

polyphyletic groups of organisms that have descended from the same ancestral species

prokaryote organism whose cells contain a single chromosome that is not housed within a nucleus; bacteria and archaea are prokaryotes

respiration general term in biology for the release of energy from the oxidation of organic molecules

ribosomes cellular structures responsible for protein synthesis

sucrose a disaccharide sugar consisting one glucose molecule bonded to a fructose molecule

sugar fungus *Saccharomyces cerevisiae*

telomere DNA sequences that protect the ends of chromosomes

transcription the formation of messenger RNA (mRNA) molecules from DNA

translation the formation of proteins by ribosomes that read mRNA molecules that specify sequences of amino acids

vaginal yeast *Candida albicans*

INDEX